SUBSTANTIVE THEORY AND CONSTRUCTIVE MEASURES

SUBSTANTIVE THEORY AND CONSTRUCTIVE MEASURES

A Collection of Chapters and
Measurement Commentary on
Causal Science

MARK STONE
—— and ——
JACK STENNER

SUBSTANTIVE THEORY AND
CONSTRUCTIVE MEASURES
A COLLECTION OF CHAPTERS AND MEASUREMENT
COMMENTARY ON CAUSAL SCIENCE

iUniverse books may be ordered through booksellers or by contacting:

iUniverse
1663 Liberty Drive
Bloomington, IN 47403
www.iuniverse.com
1-800-Authors (1-800-288-4677)

ISBN: 978-1-5320-3651-4 (sc)
ISBN: 978-1-5320-3653-8 (hc)
ISBN: 978-1-5320-3652-1 (e)

Library of Congress Control Number: 2017919668

Print information available on the last page.

iUniverse rev. date: 08/08/2018

AUTHORS

Jack Stenner and Mark Stone have been collaborating for more than forty years on issues of measurement and reading theory.

Dr. Stone is a licensed clinical psychologist, board certified in school psychology and clinical psychology (ABPP). A MESA graduate, he retired as vice president and academic dean of the Adler Institute, where he taught research design and statistics. He now teaches at Aurora University and supervises doctoral dissertations.

Dr. Stenner is chief science officer and cofounder of MetaMetrics Inc. He is research professor in the Applied Developmental Sciences and Special Education program, School of Education, at the University of North Carolina at Chapel Hill, and chief science officer at MetaMetrics— developers of the Lexile Framework for Reading and the Quantile Framework for Mathematics.

Dr. Donald Burdick, PhD, coauthored with Jack and Mark *How to Model and Test for Mechanisms That Make Measurement Systems Tick.*

Stone and Stenner propose *Substantive-Theory and Constructive Measures* as crucial elements in determining predictive measures and variance to advance causation in a specified frame of reference. The collected chapters and supplementary measurement commentary provide the details to this approach. Redundancy is purposeful in demonstrating the primacy of theory over data. The collective process is contained in the measurement mechanism that embodies substantive theory, constructed instrumentation, and assembled data supporting spot-on prediction or identifying error—causal science.

CONTENTS

ACKNOWLEDGMENTS

The authors would like to acknowledge the tireless scholarship that Kenneth Royal and Richard Smith extend to the global Rasch community via *Rasch Measurement Transactions* and the *Journal of Applied Measurement*. We appreciate their permission to reprint selected chapters and essays (boxes).

This book benefited enormously from the editorial contributions and computer technology skills of Travis Ruopp.

INTRODUCTION

Chapter 1 gives an overview of our position on making measures. The process involves integrating a substantive causal theory with Georg Rasch's logistic model. Substantive changes inform what differences or outcomes will result from connecting object measures, instrument calibrations, and a specification equation. This process is referred to throughout the book as a *measuring mechanism*.

A goal of measurement is general objectivity. Rasch's specific objectivity requires that differences or comparisons between persons be independent of the instrument. A canonical case, whereby each person is measured by a unique instrument, illustrates the extreme limit of this scenario in which no overlap exists between instruments and persons, offering a new perspective on reliability and validity. This approach aligns reading ability in the social science realm with measurement of temperature in the physics realm. The data tables in Chapter 2 illustrate the alignment of the two strategies.

Changes in maps from the past to the present offer a visual illustration of the state of pictorial knowledge as the development of this knowledge facilitates the developmental progress in mapmaking and yields continuous improvement and more accuracy. Chapter 3 gives numerous examples that serve to illustrate the developmental process in a science field where continuous refinement serves to improve precision.

How shall we explain reading ability? Does the reader comprehend the text because the reader is able or because the text is easy? We argue that both reader ability and complexity factor in understanding reading comprehension. Upon reviewing the problems of measuring reader ability and text complexity, a detailed explication in Chapter 4 of the Lexile Framework illustrates how this is accomplished. Similar to measuring temperature, a physical scale serves to parallel reading ability and allows measurement by analogy.

A causal Rasch model (CRM) involves experimental intervention/ manipulation integrated with a substantive theory. A specification equation details the causal approach as a model for measurement validity. Chapter 5 introduces the trade-off property resulting from this approach. Similar to the physics triplet $F = MA$ and other 3-variable examples, a manipulation of one variable results in predictable changes in a second variable when the third is held constant.

The explication given in chapter 6 of Rasch's specific objectivity is illustrated by his simple experiment involving various ashtrays dropped from different heights, leading to the understanding of comparison as a foundation of measuring. Rasch's essential equations are reviewed, and the implications of his experiment are discussed.

A causal Rasch model involves experimental intervention/manipulation on either reader ability or text complexity, or a conjoint intervention on both to yield a successful prediction of the resulting observed outcome (count correct). Chapter 7 explains that a substantive theory shows what interventions/manipulations to the measurement mechanism can be traded off against a change to the object measure to hold the observed outcome constant. This approach parallels the description of a well-known physics law given in box 21, the combined gas law and a Rasch reading law, and in box 15, "Specification equations in causal models."

A short commentary is given in chapter 8 on the cubit, one of history's oldest units of measurement. It may be the longest operating unit with over three thousand years in use, illustrating strong substantive theory over time.

Georg Rasch, for too long, has been associated with item analysis in contradistinction to his work in other investigations. We offer in Chapter 9 his explication and our commentary on the issue of measuring growth.

Boxes containing additional/supplemental material following each chapter or are interspersed within chapters. Some are short and focused, while others provide more information and commentary. These boxes serve to explain, define, connect, and illustrate chapter content, replacing footnotes and a glossary. Our view of validity is summarized in the last box, affirming and enhancing the position taken by Lumsden and Ross (1973) more than forty years ago. We frame this approach by showing the model for how physical scientists think and how social scientists can learn to think.

References to all chapters and boxes and the index are contained in the final two chapters. Several figures are repeated in sections of the book to keep them current with the accompanying narrative. All biblical references use the Revised Standard Version (RSV). See reference section.

HOW TO MODEL AND TEST FOR THE MECHANISMS THAT MAKE MEASUREMENT SYSTEMS TICK

Introduction

The vast majority of psychometric thought over the last century has had as its focus the item. Shortly after Spearman's original conception of reliability as whole instrument (1904), replication proved to be difficult, as there existed little understanding of what psychological instruments actually measured. The lack of substantive theory made it difficult indeed to "clone" an instrument (make a genetic copy). In the absence of a substantive theory the instrument maker does not know what features of test items are essential to copy and what features are incidental and cosmetic (Irvine and Kyllonen 2002). Faced with the need to demonstrate the reliability of psychological instruments but lacking a substantive construct theory that would support instrument cloning, early psychometrics took a fateful step inward. Spearman (1910) proposed estimating reliability as the correlation between sum scores on odd and even items of a single instrument. Thus was the instrument lost as a focus of psychometric study, and the part score and inevitably the item became ascendant. This inward misstep gave rise to thousands of instruments with nonexchangeable metrics populating a landscape devoid of unifying psychological theory. And this is so because "The route from theory or law to measurement can almost never be traveled backwards" (Kuhn 1963).

There are two quotes that when taken at extreme face value open up a new paradigm for measurement in the social sciences:

> It should be possible to omit several test questions at different levels of the scale without affecting the individual's [reader's] score [measure]. (Thurstone 1926)

1

A comparison between two individuals [readers] should be
independent of which stimuli [test questions] within the class
considered were instrumental for comparison; and it should also
be independent of which other individuals were also compared,
on the same or some other occasion. (Rasch 1961)

Both Thurstone and Rasch envisioned a measurement framework
in which individual readers could be compared independent of which
particular reading items were instrumental for the comparison. Taken to
the extreme, we can imagine a group of readers being invariantly ordered
along a scale without a single item in common (no two readers exposed
to the same item). This would presumably reflect the limit of "omitting"
items and making comparisons "independent of the items" used to make
the comparison. Compare a fully crossed data collection design (each
item is administered to every reader) with a design in which items are
nested in persons (items are unique to each person). Although easily
conceived, it is immediately clear that no existing data analysis method
can extract invariant reader comparisons from the second design type
data. But is this not exactly the kind of data that is routinely generated; for
example, when parents report their child's weight on a doctor's office form?
No two children (except for siblings) share the same bathroom scale, or
potentially even the same underlying technology, and yet we consistently
and invariantly order all children in terms of weight. What is different is
that the same construct theory for weight has been engineered into each
and every bathroom scale even though the specific mechanism (digitally
recorded pressure versus spring-driven analog recording) may vary.

In addition, the measurement unit (pounds or kilograms) has been
consistently maintained from bathroom scale to bathroom scale.

Social science measurement does not, as a rule, make use of substantive
theory in the ways that the physical sciences do.

Validity theory and practice suffers from an egalitarian malaise—all
correlations are considered part of the fabric of meaning, and like so many
threads, each is treated equally. Because we live in a correlated world,
correlations of absolute zero are rare, nonzero correlations abound, and it
is an easy task to collect a few statistically significant correlates between
scores produced by virtually any human science instrument and other
meaningful phenomena. All that is needed to complete our validity tale is
a story about why so many phenomena are correlated with the instrument
we are making.

And so it goes, countless times per decade: dozens of new instruments
are islands unto themselves accompanied by hints of connectivity

whispered through dozens of middling correlations. This is the legacy of the nomological network (Cronbach and Meehl 1955). May it rest in peace!

Validity, for us, is a simple, straightforward concept with a narrow focus.

It answers the question, "What causes the variation detected by the instrument?" The instrument (a reading test) by design comes in contact with an object of measurement (a reader), and what is recorded is a measurement outcome (count correct). That count is then converted into a linear quantity (a reading ability). Why did we observe that particular count correct? What caused a count correct of 25/40 rather than 20/40 or 30/40? The answer (always provisional) takes the form of a specification equation with variables that when experimentally manipulated produce the changes in item behavior (empirical item difficulties) predicted by the theory (Stenner, Smith, and Burdick 1983). In this view validity is not about correlations or about graphical depictions of empirical item orderings called Wright maps (Wilson 2004). It is about asking, "What is causing what?" Is the construct well enough understood that its causal action can be specified? Clearly our expectation is unambiguous. There exist features of the stimuli (test or survey items) that if manipulated will cause changes in what the instrument records (what we observe). These features of the stimuli interact with the examinee, and the instrument records the interaction (correct answer, strong agreement, tastes good, etc.). The window onto the interaction between examinee and instrument is clouded. We can't observe directly what goes on in the mind of the examinee, but we can dissect and otherwise manipulate the item stimuli, or measurement mechanism, and observe changes in recorded behavior of the examinee (Stenner, Stone, and Burdick 2009). Some of the changes we make to the items will matter to examinees (radicals), and others will not (incidentals). Sorting out radicals (causes) from incidentals is the hard work of establishing the validity of an instrument (Irvine and Kyllonen 2002). The specification equation is an instantiation of these causes (at best) or their proxies (at a minimum). Typical applications of Rasch models and IRT models to human science data are thin on substantive theory. Rarely is there an a priori specification of the item calibrations (i.e., constrained models).

Instead the analyst estimates both person parameters and item parameters from the same data set. For Kuhn, this practice is at odds with the function of measurement in the "hard" sciences in that almost never will substantive theory be revealed from measurement (Kuhn 1961). Rather, according to Kuhn, "the scientist often seems rather to be struggling with facts [e.g., raw scores], trying to force them to conformity with a theory he does not doubt" (Kuhn 1961). Here Kuhn is talking about substantive theory, not axioms. The scientist imagines a world, formalizes these imaginings as a theory, and then makes measurements and checks for congruence between what is observed and what theory predicted: "Quantitative facts cease to seem simply the 'given'. They must be fought for and with, and in this fight the theory with which they are to be compared proves the most potent weapon." It's not just that unconstrained models are less potent but that they fail to conform to the way science is practiced, and most troubling, they are least revealing of anomalies (Andrich 2004).

Andrich (2004) makes the case that Rasch models are powerful tools precisely because they are prescriptive rather than descriptive, and when model prescriptions meet data, anomalies arise. Rasch models invert the traditional statistical data-model relationship by stating a set of requirements that data must meet if those data are to be useful in making measurements. These model requirements are independent of the data. It does not matter if the data are bar presses, counts correct on a reading test, or wine taste preferences: if these data are to be useful in making measures of rat perseverance, reading ability, or vintage quality, all three sets of data must conform to the same invariance requirements. When data sets fail to meet the invariance requirements, we do not respond by relaxing the invariance requirements through addition of an item specific discrimination parameter to improve fit; rather, we examine the observation model and imagine changes to that model that would bring the data into conformity with the Rasch model requirements.

A causal Rasch model (item calibrations come from theory, not the data) is doubly prescriptive (Stenner, Stone, and Burdick 2009; Stenner et al. Fisher, Stone, and Burdick 2013).

First, it is prescriptive regarding the data structures that must be present:

> The comparison between two stimuli [text passages] should be independent of which particular individuals [readers] were instrumental for the comparison; and it should also be independent of which other stimuli within the considered class [prose] were or might also have been compared. Symmetrically, a comparison between two individuals [readers] should be independent of which particular stimuli within the class considered [prose] were instrumental for [text passage] comparison; and it should also be independent of which other individuals were also compared, on the same or on some other occasion. (Rasch 1961)

Second, causal Rasch models (CRM) prescribe that item calibrations take the values imposed by the substantive theory (Burdick, Stone, and Stenner 2006; Stenner, Burdick, and Stone 2008).

Thus, the data, to be useful in making measures, must conform to both Rasch model invariance requirements and substantive theory invariance requirements as represented in the theoretical item calibrations. When data meet both sets of requirements, then those data are useful not just for making measures of some construct but for making measures of that precise construct specified by the equation that produced the theoretical item calibrations. We note again that these dual invariance requirements come into stark relief in the extreme case of "no connectivity across stimuli or examinees." How, for example, are two readers to be measured on the same scale if they share no common text passages or items? If you read a Harry Potter novel and I read *Lord of the Rings*, how is it possible that from these disparate experiences an invariant comparison of our reading abilities is realizable? How is it possible that you can be found to read 250L better than I can, and furthermore that you had 95 percent comprehension and I had 75 percent comprehension of our respective books? Given that seemingly nothing is in common between the two experiences, it seems that invariant comparisons are impossible, but recall our bathroom scale example: different instruments qua experiences underlie every child's parent-reported weight. Why are we so quick to accept that you weigh fifty pounds less than I do and yet find claims about our relative reading abilities (based on measurements from two different books) inexplicable? The answer lies in well-developed construct theory, instrument engineering, and metrological conventions.

Clearly, each of us has had ample confirmation that the construct "weight" denominated in pounds or kilograms can be well measured by any well-calibrated bathroom scale. Experience with diverse bathroom scales has convinced us that within a pound or two of error these instruments will produce not just invariant relative differences between two persons (as described in the Rasch quotes) but the more stringent expectation of invariant absolute magnitudes for each individual independent of instrument.

Over centuries, instrument engineering has steadily improved to the point that for most purposes "uncertainty of measurement" (usually reported as the standard deviation of a distribution of imagined or actual replications taken on a single person) can be effectively ignored for most bathroom scale applications. Finally, by convention (i.e., the written or unwritten practice of a community) in the United States we denominate weight in pounds and ounces. The use of pounds and ounces is arbitrary, as is evident from the fact that most of the world has gone metric, but what is decisive is that a unit is agreed to by the community and is slavishly maintained through consistent implementation, instrument manufacture, and reporting. At present, "reading ability" does not enjoy a commonly adhered to construct definition, nor a widely promulgated set of instrument specifications, nor a conventionally accepted unit of measurement, although the Lexile Framework for Reading promises to unify the measurement of reading ability in a manner precisely parallel to the way unification was achieved for length, temperature, weight, and dozens of other useful attributes (Stenner, Burdick, Sanford, and Burdick 2006; Stenner and Stone 2010).

A causal (constrained) Rasch model that fuses a substantive theory to a set of axioms for conjoint additive measurement affords a much richer context for the identification and interpretation of anomalies than does an unconstrained Rasch model (Stenner, Stone, and Burdick 2009). First, with the measurement model and the substantive theory fixed, it is self-evident that anomalies are to be understood as problems with the data, ideally leading to improved observation models that reduce unintended dependencies in the data. Recall that the duke of Tuscany put a top on some of the early thermometers, thus reducing the contaminating influences of barometric pressure on the measurement of temperature. He did not propose parameterizing barometric pressure so that the boiling point of water at sea level would match the model expectations at three thousand feet above sea level. Second, with both model and construct theory fixed, it is obvious that our task is to produce measurement outcomes that fit the (aforementioned) dual invariance requirements. By analogy, not all fluids are ideal as thermometric fluids. Water, for example, is nonmonotonic in its expansion with increasing temperature. Mercury, in contrast, has many useful properties as a thermometric fluid. But the discovery that not all fluids are useful thermometric fluids does not invalidate the concept of temperature. Rather, the existence of a single fluid with the necessary properties validates temperature as a useful construct. The existence of a persistent invariant framework makes it possible to identify anomalous behavior (i.e., the interaction of water and barometric pressure) and interpret it in an expanded theoretical framework. Analogously, finding that not all reading item types conform to the dual invariance requirements of a Rasch model and the Lexile theory does not invalidate either the axioms of conjoint measurement theory or the Lexile reading theory. Rather, anomalous behaviors of various item types are open invitations to expand the theory to account for these deviations from expectation. Notice here the subtle shift in perspective. We do not need to find one thousand unicorns; a single one will establish the reality of the class.

The finding that reader behavior on a single class of reading tasks can be regularized by the joint actions of the Lexile theory and a Rasch model is sufficient evidence for the reality of the reading construct (Michell, 1999).

Model and Theory

Equation (1) is a causal Rasch model for dichotomous data, which sets a measurement outcome (raw score) equal to a sum of modeled probabilities:

$$Expected\ raw\ score =: \sum_i \frac{e^{(b-di)}}{1 + e^{(b-di)}}$$

The measurement outcome is the dependent variable, and the measure (e.g., person parameter, b) and instrument (e.g., the parameters di pertaining to the difficulty d of item i) are independent variables. The measurement outcome (e.g., count correct on a reading test) is observed, whereas the measure and instrument parameters are not observed but can be estimated from the response data and substantive theory, respectively. When an interpretation invoking a predictive mechanism is imposed on the equation, the right-side variables are presumed to characterize the process that generates the measurement outcome on the left side. The symbol =: was proposed by Euler circa 1734 to distinguish an algebraic identity from a causal identity (right-hand side causes the left-hand side). The same symbol =:, later exhumed by Judea Pearl, can be read as *manipulation of the right-hand side via experimental intervention will cause the prescribed change in the left-hand side of the equation.*

A Rasch model combined with a substantive theory embodied in a specification equation provides a more or less complete explanation of how a measurement instrument works (Stenner, Stone, and Burdick 2009). In the absence of a specified measurement mechanism, a Rasch model is merely a probability model. A probability model absent a theory may be useful for describing or summarizing a body of data and for predicting the left side of the equation from the right side, but a Rasch model in which instrument calibrations come from a substantive theory that specifies how the instrument works is a causal model. That is, it enables prediction after intervention (Woodward 2003).

Causal models (assuming they are valid) are much more informative than probability models: "A joint distribution tells us how probable events are and how probabilities would change with subsequent observations, but a causal model also tells us how these probabilities would change as a result of external interventions ... Such changes cannot be deduced from a joint distribution, even if fully specified" (Pearl [2000] 2009).

A satisfying answer to the question of how an instrument works depends on understanding how to make changes that produce expected effects. Two identically structured examples of such narratives are a thermometer designed to take human temperature and a reading test.

1.1. The NexTemp Thermometer

The NexTemp thermometer is a small plastic strip pocked with multiple enclosed cavities. In the Fahrenheit version, forty-five cavities arranged in a double matrix serve as the functioning end of the unit. Spaced at 0.2°F intervals, the cavities cover a range from 96.0°F to 104.8°F. Each cavity contains three cholesteric liquid crystal compounds and a soluble additive. Together, this chemical composition provides discrete and repeatable change-of-state temperatures consistent with the device's numeric indicators. Change of state is displayed optically and is easily read.

1.2. The Lexile Framework for Reading

Text complexity is predicted from a construct specification equation incorporating sentence length and word frequency components. The squared correlation of observed and predicted item calibrations across hundreds of tests and millions of students over the last 15 years averages about $R2 = 0.93$.

Available technology for measuring reading ability employs computer-generated items built on the fly for any continuous prose text. Counts correct are converted into Lexile measures via a Rasch model estimation algorithm employing theory-based calibrations. The Lexile measure of the target text and the expected spread of the cloze items are given by theory and associated equations. Differences between two readers' measures can be traded off for a difference in Lexile text measures. When the item generation protocol is uniformly applied, the only active ingredient in the measurement mechanism is the choice of text and its associated complexity.

In the temperature example, if we uniformly increase or decrease the amount of soluble additive in each cavity, we change the correspondence table that links the number of cavities that turn black to degrees Fahrenheit. Similarly, if we increase or decrease the text demand (Lexile) of the passages used to build reading tests, we predictably alter the correspondence table that links count correct to Lexile reader measure. In the former case, a temperature theory that works in cooperation with a Guttman model produces temperature measures. In the latter case, a reading theory that works in cooperation with a Rasch model produces reader measures. In both cases, the measurement mechanism is well understood, and we exploit this understanding to address a vast array of counterfactuals (Woodward 2003). If things had been different (with the instrument or object of measurement), we could still answer the question as to what then would have happened to what we observe (i.e., the measurement outcome). It is this kind of relation that illustrates the meaning of the expression "there is nothing so practical as a good theory" (Lewin 1951).

Distinguishing Features of Causal Rasch Models

Clearly the measurement model we have proposed for human sciences mimics key features of physical science measurement theory and practice. Below we highlight several such features.

1. The model is individual centered. The focus is on explaining variation within person over time.

Much has been written about the disadvantages of studying between-person variation with the intent to understand within-person causal mechanisms (Grice 2011; Barlow, Nock, and Hersen 2009). Molenaar (2004) has proven that only under severely restrictive conditions can such cross-level inferences be sustained. In general in the human sciences, we must build and test individual-centered models and not rely on variable or group-centered models (with attendant focus on between-person variation) to inform our understanding of causal mechanisms. Causal Rasch models are individually centered measurement models. The measurement mechanism that transmits variation in the attribute (within person over time) to the measurement outcome (count correct on a reading test) is hypothesized to function the same way for every person (the second ergodicity condition of homogeneity) (Molenaar 2004). Note, however, that the existence of different developmental pathways that led you to be taller than me and me to be a better reader than you does not mean that the attributes of height and reading ability are necessarily different attributes for both of us.

2. In this framework, the measurement mechanism is well specified and can be manipulated to produce predictable changes in measurement outcomes (e.g., percent correct).

For purposes of measurement theory, we don't need a sophisticated philosophy of causal inference. For example, questions about the role of human agency in the intervention/manipulation-based accounts of causal inference are not troublesome here. All we mean by the claim that the right-hand side of equation 1 causes the left-hand side is that experimental manipulation of each will have a predictable consequence for the measurement outcome (expected raw score). Stated more generally, what we mean by "x causes y" is that an intervention on x yields a predictable change in y. The specification equation used to calibrate instrument/items is a recipe for altering just those features of the instrument/items that are causally implicated in the measurement outcome. We term this collection of causally relevant instrument features the "measurement mechanism." It is the "measurement mechanism" that transmits variation in the attribute (e.g., temperature, reading ability) to the measurement outcome (number of cavities that turn black or number of reading items answered correctly).

Two additional applications of the specification equation are: (1) the maintenance of the unit of measurement independent of any particular instrument or collection of instruments, and (2) bringing nontest behaviors (reading a Harry Potter novel, 980L) into the measurement frame of reference (Stenner and Burdick 2011).

3. Item parameters are supplied by substantive theory, and, thus, person parameter estimates are generated without reference to or use of any data on other persons or populations.

It is a feature of the Rasch model that differences between person parameters are invariant to changes in item parameters, and differences between item parameters are invariant to change in person parameters. These invariances are necessarily expressed in terms of differences because of the one degree of freedom over parameterization of the Rasch model (i.e., locational indeterminacy). There is no locational indeterminacy in a causal Rasch model in which item parameters have been specified by theory.

4. The quantitivity hypothesis can be experimentally tested by evaluating the trade-off property for the individual case (Michell 1999). A change in the person parameter can be offset or traded off for a compensating change in text complexity to hold comprehension constant. The trade-off is not just about the algebra in equation 1. It is about the consequences of simultaneous intervention on the attribute (reader ability) and measurement mechanism (text complexity). Careful thinking about quantitivity makes the distinction between "an attribute" and "an attribute

as measured." The attribute "hardness" is not quantitative as measured on the Mohs scale, but it is quantitative as measured on the Vickers scale (1923). So it is confusing to talk about whether an attribute, in and of itself, is quantitative or not. If an attribute "as measured" is quantitative, then it can always be represented as merely ordinal. But the obverse is not true. Twenty-first-century science still uses the Mohs scratch test, which produces more-than and less-than statements about the "hardness" of materials. Pre-1923, it would have been inaccurate to claim that hardness "as measured" was a quantitative attribute, because no measurement procedure had yet been invented that produced meaningful differences (the Mohs scratch test produces meaningful orders but not meaningful differences). The idea of dropping, with a specified force, a small hammer on a material and measuring the volume of the resulting indentation opened the door to testing the quantitivity hypothesis for the attribute "hardness." "Hardness" as measured by the falling hammer passed the test for quantitivity, and correspondence tables now exist for reexpressing mere order (Mohs) as quantity (Vickers). Michel (1999) states:

> Because measurement involves a commitment to the existence of quantitative attributes, quantification entails an empirical issue: is the attribute involved really quantitative or not? If it is, then quantification can sensibly proceed. If it is not, then attempts at quantification are misguided. A science that aspires to be quantitative will ignore this fact at its peril. It is pointless to invest energies and resources in an enterprise of quantification if the attribute involved is not really quantitative. The logically prior task in this enterprise is that of addressing this empirical issue. (75)

As we have just seen, we cannot know whether an attribute is quantitative independent of attempts to measure it. If Vickers's company had Michel's book available to them in 1923, then they would have looked at the ordinal data produced by the Mohs scratch test and concluded that the "hardness" attribute was not quantitative, and thus it would have been "misguided" and "wasteful" to pursue his hammer test. Instead Vickers and his contemporaries dared to imagine that "hardness" could be measured by the hammer test and went on to confirm that "hardness as measured" was quantitative.

Successful point predictions under intervention necessitate quantitative predictors and outcomes. Concretely, if an intervention on the measurement mechanism (e.g., an increase in the text complexity of a reading passage by 250L) results in an accurate prediction of the measurement outcome (e.g., how many reading items the reader will answer correctly), and if this process can be successfully repeated up and down the scale, then text complexity, reader ability, and comprehension (success rate) are quantitative attributes of the text, person, and reader/text encounter, respectively. Note that if text complexity was measured on an ordinal scale (think Mohs), then making successful point predictions about counts correct based on a reader/text difference would be impossible. Specifically, successful prediction from differences requires that what is being differenced has the same meaning up and down the respective scales. Differences on an ordinal scale will lead to inconsistent predictions precisely because "one more" means something different depending on where you are on the scale.

Note that in the Rasch model, performance (count correct) is a function of an exponentiated difference between a person parameter and an instrument (item) parameter. In the Lexile Framework for Reading (LF), equation 1 is interpreted as:

Comprehension = Reader Ability – Text Complexity
(success rate)

The algebra in equation 1 dictates that a change in reader ability can be traded off for an equal change in text complexity to hold comprehension constant. However, testing the "quantivity hypothesis" requires more than the algebraic equivalence in a Rasch model. What is required is an experimental intervention/manipulation on either reader ability or text complexity or a conjoint intervention on both simultaneously that yields a successful prediction on the resultant measurement outcome (count correct). When manipulations of the sort just described are introduced for individual reader/text encounters and model predictions are consistent with what is observed, the quantivity hypothesis is sustained. We emphasize that the above account is individual centered as opposed to group centered. The LF purports to provide a causal model for what transpires when a reader reads a text. Nothing in the model precludes averaging over readers and texts to summarize evidence for the "quantivity hypothesis," but the model can be tested at the individual level. So just as pressure and volume can be traded off to hold temperature constant or volume and density can be traded off to hold mass constant, reader ability and text complexity can be traded off to hold comprehension constant. Following Michell (1999), we note that a trade-off between equal increases (or decrements) in text complexity and reader ability "identifies equal ratios directly" and that "identifying ratios directly via trade-offs results in the identification of multiplicative laws between quantitative attributes. This fact connects the theory of conjoint measurement with what Campbell called derived measurement" (Michell 1999).

Garden-variety Rasch models and IRT models are in their application purely descriptive. They become causal and law-like when manipulations of the putative quantitative attributes produce changes (or not) in the measurement outcomes that are consistent with model predictions. If a fourth-grade reader grows 100L in reading ability over one year and the text complexity of her fifth grade science textbook also increases by 100L over the fourth-grade-year textbook, then the forecasted comprehension rate (whether 60 percent, 70 percent, or 90 percent) that that reader will enjoy in fifth-grade science remains unchanged. Only if reader ability and text complexity are quantitative attributes will experimental findings coincide with these model predictions. We have tested several thousand students' comprehension of 719 articles averaging 1,150 words. Total reading time was 9,794 hours, and the total number of unique machine-generated comprehension items was 1,349,608. The theory-based expectation was 74.53 percent correct, and the observed outcome was 74.27 percent correct.

Conclusion

This chapter has considered the distinction between a descriptive Rasch model and a causal Rasch model. We have argued for the importance of measurement mechanisms and specification equations. The measurement model proposed and illustrated (using NexTemp thermometers and the Lexile Framework for Reading) mimics in several important ways physical science measurement theory and practice. We plead guilty to "aping" the physical sciences and despite the protestations of Michell (1999) and Markus and Borsboom (2011) do not view as tenable any of the competing go-forward strategies for the field of human science measurement.

Box 1: On Temperature

The 1560 print *Temperance* by Pieter Bruegel the Elder (see below) illustrates "pantometry" in action. Each section of this print illustrates a practical application of the mathematical sciences and measurement. These scenes illustrate quantification attempts across many aspects of measuring; use of a divider, square, plumb line, visual sighting, or aspects of velocity/distance with cannons or crossbow together with disputation also serve a prominent role.

Quantification and visualization go hand in hand with observations by providing the key to understanding measurement. Arithmetic, geometry, and trigonometry share with writing and music the pursuit of uniform quanta. Writing and music are linear events that embody the principle of measurement no less than any other area of science. Bruegel captured more than just the historical scene; he pictures the essence of metrology—a continuous search for units with generality.

Applicability and usefulness of units requires that all measures (and units) possess sensus communis, or "common sense" as Kant ([1798] 1917) expressed it. Kant meant that communication among peoples is not possible without a "common sense" operating. Visualizing measurement is applying common sense by the use of pictures, graphs, maps, and so on. This approach is the key to success in communication, utility, and generality (Stone, Wright, and Stenner 1999). Measurement is always made by means of an analogy. Hans Vaihinger (1924) wrote:

> All cognition is the apperception of one thing through another … we are always dealing with an analogy and we cannot imagine how otherwise existence can be understood … all knowledge can only be analogical. (29)

Common examples from the past for measuring time include the tolling of bells, sundials, and water clocks. Today we have digital watches and atomic clocks for measuring time with greater accuracy. We say, "Time passes," or, "Time marches on," and record the duration in terms of length. There is no "time," only duration. Length is the analogy for duration. A theory of time as duration is transformed by analogy from a variable of length and made manifest using natural occurrences such as the sun, moon, and stars and artificial devices as mentioned earlier. Robert Oppenheimer (1955) in his address to the American Psychological Association entitled *Analogy in Science* said:

> Whether or not we talk of discovery or of invention, analogy is inevitable in human thought, because we come to new things in science with what equipment we have, which is how we have learned to think, and above all how

we have learned to think about the relatedness of things. We cannot, coming into something new, deal with it except on the basis of the familiar and the old fashioned. ... We cannot learn to be surprised or astonished at something unless we have a view of how it ought to be; and that view is almost certainly an analogy. (129–130)

Rasch (1961) addressed this problem with a theory, a class of models, and specific data examples. His goal was "replacing qualitative observations by quantitative parameters" (331).

Consider temperature and common measurement. Temperature for most of us means the heat or cold we experience in our environment. In day-to-day life as well as in the more rigorous conditions of scientific laboratories, temperature requires some analogous method by which to make measures. A thermometer commonly uses an expansion tube of mercury to accomplish this task. Water and alcohol were among other elements investigated in arriving at the choice of mercury. Variations abound on the way to utility.

Celsius and Fahrenheit Temperature

The common indoor/outdoor thermometer usually shows both Celsius and Fahrenheit:

Figure B1.1 Temperance

For practical purposes, the thermometer is simply an "expansion tube" of mercury. The elevation (length) of mercury in the tube is analogous to temperature. This elevation is made utilitarian via the association of numerals to our personal sensations of comfort/discomfort. Thirty degrees F is experienced as cold, and seventy degrees F is considered warm. In countries using Celsius, zero degrees C and twenty degrees C convey approximately the same sensations. The two scales, C and F, illustrated in the figure are not different. The distances between the two horizontal lines indicating high and low (F_h and F_l, or C_h and C_l) show an equal vertical distance of length on the F and C scales. Any line drawn horizontally across the two tubes will indicate exactly the same elevation on both scales.

One intriguing aspect of this instrument is that volume in three dimensions for the thermometer has been reduced to length in one dimension for interpreting temperature. A complex variable has been reduced to a simple one. Rasch ([1960] 1980) in discussing models in classical physics remarked,

> None the less it should not be overlooked that the laws do not at all give an accurate picture of nature. They are simplified descriptions of a very complicated reality. (10, our emphasis)

Judging from the voluminous amount of commentary in the social sciences citing how "complicated" reality is and how difficult it is to model, this point seems rarely appreciated. Physics has progressed admirably following "simple" laws to model complex matters. Scientists appreciate complexity, but nature cannot be understood when complexity is made a stumbling block to understanding. In such instances, emphasizing complexity obfuscates understanding and knowledge. This temperature example reminds us that complexity can be modeled in a simple fashion if only we can find a useful way to do so.

Celsius and Fahrenheit report different temperature numerals but not different temperatures. What is different between these two temperature scales is their *division* of length into segments, each one with different units and different origins for locating zero degrees. It is the numerals that differ, not the temperature, because the values of C and F can be connected by the algebraic expression $9C = 5F - 160$. Entering C and solving for F, or vice versa, gives us the corresponding value.

A horizontal line across the picture of the temperature tubes supplies all the visual analogy we need to move from one scale to the other. This is because the "height" (i.e., the length of mercury in the tube) is invariant. It is the same height for each scale. The C and F scales are shown to be equal by observing this line connecting the two lengths. Algebra connects these two different scales precisely. What is not the same are the respective scale divisions, and there are numerous variants.

This simple example has significant implications for understanding the essence of measuring:

1. We measure by analogy, whether with moving hands, ticking clock, sand trickling through an hourglass, or cesium clock. No matter how sophisticated the device, analogy prevails in some form.
   For temperature: mercury is a visual representation on the quantitative scale(s).

2. We should not be confused by differing scale values and origins into thinking complexity abounds. A validly constructed instrument emanates from a single, unified variable. The problem is to devise and construct one.
   For temperature: there is only one construct variable but many ways to divide and express it.

3. Validity rests on achieving instrument integrity and invariance. Everything else is peripheral to this problem and only serves to confuse the matter. Constructing the instrument and applying it to life are two entirely different matters not to be confused.
   For temperature: the instrument is foundational; applications follow.

4. Portability is necessary. Handled properly, the instrument is useful in almost all locations. Extreme conditions in temperature or elevation above/below sea level require modifications and corresponding interpretation.
   For temperature: general application and utility constitute validity with some unique exceptions.

5. Utility is an important aspect of measuring. Given the choice between two explanations, complex versus simple, Occam's razor favors the simple as the useful one. Utility implies understanding.
   For temperature: giving one's attention to observing the temperature, and not to the instrument, illustrates the successful achievement of utility.

Box 2: Why the Thermometer Is Such a
Good Paradigm for Measurement

Is any aspect of our environment more all-encompassing than temperature? The fact that we usually take it for granted, except for unusual events, indicates how pervasive we feel its influence to be. There is a long history of temperature investigation and development, which continues to the present day. It encompasses both the everyday use of practical instruments as well as scientific investigations, especially of extreme heat and cold. Everyday life exposes us to temperature conditions and thermometer results directly (in our own life) or indirectly (mediated by TV, newspaper, etc.) Temperature involves not just the environment around us but also the human condition ("Do I have a fever?").

What is temperature? We don't measure it directly but analogously, previously by tubes of water or alcohol, and now by mercury. Specialized devices assess extreme ranges of heat and cold, but for most of us, the common mercury thermometer providing a Celsius or Fahrenheit measure is common.

Many physical effects have been used as indirect indices of temperature change. We have seen the development from the thermoscope to the thermometer as one line of improvement. Science has also come to learn how pressure, volume, and temperature relate to one another (Burdick, Stone, and Stenner 2006).

Temperature is an index of heat, one form of energy determined by inflows and outflows. Thermodynamics is concerned with the relationship of energy in motion, atoms and molecules, and their movement in matter. Motion energy, kinetic energy, increases in proportion to the square of the velocity of moving objects.

Thermometric devices take the averages of the kinetic energy of voluminous measures (the ensemble). The kelvin K is the accepted unit of (theoretical) thermodynamic temperature with unit differences that are exactly equal. This practical approach is the consequence of the International Temperature Scale of 1990 (ITS-90), earlier called the International Practical Temperature Scale. The Kelvin scale is anchored at absolute zero with ITS-90 establishing thirteen calibration points by elements, of which five are spaced below the triple point of water (a compound) and eight are spaced above. Examples include the triple point of hydrogen (13.8033 K), triple point of oxygen (54.3584 K), triple point of water (273.16 K), freezing point of zinc (692.677 K), and the freezing point of copper (1357.77 K).

The mission of ITS-90, as for all previous committees, was to align the practical, an approximation, to the thermodynamic temperature scale, a theoretically ideal scale. They traditionally provide updated tables of their achievements.

In essence, the story of thermonuclear temperature metrology supports (1) the primacy of theory (the thermonuclear scale) as a driver of practical achievement in physics; and (2) the role of instrument construction and data in the history of theory. These are the essential considerations for inventing measures.

GENERALLY OBJECTIVE MEASUREMENT OF HUMAN TEMPERATURE AND READING ABILITY

2

In this paper, we look closely at two latent variables (temperature and reading) and two instruments used in their measurement. At first glance, the conceptual foundations for these constructs seem to represent quite different entities. A thermometer measures human temperature widely assumed to be a physical attribute, while a test for measuring reader ability clearly represents a mental attribute. Common sense tends to assert fundamentally dissimilar ontologies. We intend to show that underneath this surface dissimilarity, there are striking parallels that can be exploited to illuminate the "oughts" of human science measurement. Moreover, our expectation is that this comparison will offer important insights to health outcome researchers by showing the efficiency of linear measurement for establishing closed knowledge systems. Closed knowledge systems offer important benefits to effective rehabilitation by isolating key constructs affecting patient functional status. Their parameterization in a closed system diminishes treatment uncertainty and increases measurement efficiency. Both thermodynamics and Lexile theory emphasize latent constructs that are operationally defined not only by units but comprehensive "deep structure" substantive theories, which guide general system formulation. Both thermodynamics and Lexile theory are systems that manipulate only a few key constructs, which are applicable across broad classes of physical and mental activity. This approach to latent traits suggests health outcome measurement is also probably defined by several deep structure constructs that permeate outcome measurement. Closed systems conforming to substantive theories that follow thermodynamics and Lexile models can likely be developed to govern these constructs.

Before comparing and contrasting temperature and reader ability, we assert the physical science constructs *temperature* and the human science construct *reader ability* share common philosophical foundations. Both latent constructs signify real entities whose causal action can be manipulated and their effects observed on interval scales. Neither construct should be conceived of as "just a useful fiction." Thus, we reject a constructivist interpretation for either construct. Both constructs are attributes of human beings, and we are entity realists. The relationship between the latent variable and the measurement outcome is causal at both the intraindividual and interindividual, stated more formally, the conditional probability distribution of the measurement outcome given the latent variable is to be given a *stochastic subject interpretation* (Holland 1990). We reject a repeated sampling interpretation of the conditional probability distribution. We assert for both temperature and reading ability, the measurement takes the same form within and between persons, what Ellis and Van den Wollenberg (1993) called the *local homogeneity assumption.* We assert this assumption is testable in the theory referenced measurement context (Stenner, Smith, and Burdick 1983).

Figure 1 presents an aspect chart for two latent variables: temperature and reader ability. For our purposes, we assume that the object of measurement in both cases is a person. In each case, the instrument (thermometer or reading test) is brought into contact with the person. A measurement outcome (number of cavities (0–45) that fail to reflect green light or count correct on forty-five theoretically calibrated reading items) is recorded by a professional (nurse or teacher). Note that both instruments have been targeted on the appropriate range for each person. The thermometer measures from 96° to 104.8°F, and let's assume the reading test has been targeted for a typical fourth grader (500L to 900L). The substantive theory provides the link between the measurement outcome and the measure denominated in a conventional unit (degrees Fahrenheit, °F, or Lexiles, L). Note that substantive theory is used to build the respective instrument and to convert cavity counts and counts correct on the reading test into a measure. Cavity counts and counts correct are sufficient statistics for their respective parameters (i.e., measure). Sufficient statistics exhaust the information in the data that is relevant to estimate the parameter/measure. In each case, a point estimate of the measure is produced for the person without recourse to information about other persons' measures. Instrument calibrations come from theory, and individual instruments may be disposable—have never been used before and will never be used again. We turn now to a more detailed look at the NexTemp thermometer and MyReadingWeb to further draw out the parallel structures underlying of these latent variables.

Human Body Temperature

Temperature is a physical property of systems, including the human system that imperfectly corresponds to the human sensation of hot and cold. Temperature as a latent variable is a principal parameter in thermodynamic theory. Like other latent variables, temperature cannot be directly observed, but its effects can be, and thermometers are instruments that detect these effects. A thermometer has two key components: the sensor (e.g., cavities of fluid that differentially reflect light), which detects changes in temperature, and a correspondence table that converts the measurement outcome (cavity count) into a scale value (degrees Celsius) via theory (thermodynamic). A wide range of so-called primary thermometers have been built, each relying on radically physical effects (electrical resistance, expansion coefficients of two metals, velocity of sound in a monatomic gas, and gamma ray emission in a magnetic field). For primary thermometers, the relationship between a measurement outcome and its measure is so well understood that temperature readings can be computed directly from the measurement outcomes. So-called secondary thermometers (e.g., mercury thermometers) produce measurement outcomes whose relationship to temperature is not yet so well understood that temperature readings can be directly computed. In these by far more common cases, secondary thermometer readings are often calibrated against a primary thermometer, and a correspondence table is generated that links the measurement outcome to temperature expressed in, say, degrees Celsius. Celsius measures can then be converted directly into Kelvin, Fahrenheit, Rankine, Delisle, Newton, or other metrics that find use in, for example, special engineering applications or high-energy physics.

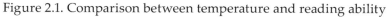

Figure 2.1. Comparison between temperature and reading ability

In 1861, Carl Wunderlich reported in a study of one million persons that the average human body temperature was 37.0 degrees Celsius. This value converts precisely to 98.6 degrees Fahrenheit and is believed to be the source for the putative "normal" temperature. Mackowiak, Wasserman, and Levine (1997) measured 148 healthy men and women multiple times each day for three consecutive days with electronic oral thermometers. The authors found that the average temperature was 36.8 degrees Celsius, which converts to 98.2 degrees Fahrenheit. They speculate that Wunderlich rounded his "average" up to 37 degrees Celsius and then the rounded measure was converted to 98.6 degrees Fahrenheit and subsequently popularized. Mackowiak, Wasserman, and Levine recommend that the popular value of 98.6 Fahrenheit for oral temperature be abandoned in favor of the new value of 98.2 and that "normal" ranges be similarly revised.

Figure 2.1 provides a black line sketch of a NexTemp disposable thermometer for measuring human temperature. The NexTemp thermometer is a thin, flexible, paddle-shape plastic strip containing multiple cavities. In the Fahrenheit version, the forty-five cavities are arranged in a double matrix at the functioning end of the unit. The columns are spaced at 0.2°F intervals covering the range of 96.0°F to 104.8°F. Each cavity contains a chemical composition comprised of three cholesteric liquid crystal compounds and a varying concentration of a soluble additive. These chemical compositions have discreet and repeatable change-of-state properties, the temperatures of which are determined by the concentrations of the additive. Additive concentrations are varied in accordance with an empirically established formula to produce a series of change-of-state temperatures consistent with the indicated temperature points on the device. The chemicals are fully encapsulated by a clear polymeric film, which allows observation of the physical change but prevents any user contact with the chemicals. When the thermometer is placed in an environment within its measure range, such as 98.2°F (37.0°C), the chemicals in all the cavities up to and including 98.2°F (37.0°C) change from a liquid crystal to an isotropic clear liquid state. This change of state is accompanied by an optical change that is easily viewed by a user. The green component of white light is reflected from the liquid crystal state but is transmitted through the isotropic liquid state and absorbed by the black background. As a result, those cavities containing compositions with threshold temperatures up to and including 98.2°F (37.0°C) appear black, whereas those with transition temperatures of 98.2°F (37.0°C) and higher continue to appear green (Medical Indicators 2006, 1–2).

In vitro accuracy of the NexTemp liquid crystal thermometer equals or exceeds glass-mercury and electronic thermometers. More than one hundred million production units have shown agreement with calibrated water baths to within 0.2°F in the range of 98.0–102.0°F and within 0.4°F elsewhere (0.1°C in the range of 37.0–39.0°C and within 0.2°C elsewhere).

In vivo tests in the United States, Japan, and Italy resulted in excellent agreement with measurements using specially calibrated glass-mercury thermometers. The mean difference between the NexTemp thermometers and the calibrated glass-mercury equilibrium device was only 0.12°F (0.07°C). The NexTemp thermometer also achieves equilibrium very rapidly, due to its small "drawdown" (cooling effect on tissue upon introduction of a room-temperature device) and the small amount of energy required to make the physical phase transition.

A competitive marketing analysis reported the following advantages of this new technology over the market-dominant older mercury in a tube technology. These "technical advantages" will prove useful in our comparison and contrast with reading test technology:

- *Cost*—The NexTemp temperature measurement technology provides lower costs when compared to other temperature devices.
- *Safety*—The safety advantages of NexTemp technology are substantial. There is no danger, as with a conventional thermometer, of glass ingestion or mercury poisoning if a child bites the active part of the unit. NexTemp and its packaging are latex-free.
- *Speed and ease-of-use*—The NexTemp thermometer is quick, portable, nonbreakable, and easy to use (e.g., no shakedown or resetting).
- *Reduced chance of cross-contamination of patients*—The NexTemp disposable product comes individually wrapped and is intended to be used and then discarded. The reusable version of the product is for single patient use over time with cleaning between uses (Medical Indicators 2006, 4).

Reader Ability

Of approximately six thousand spoken languages in the world, only about two hundred are written, and many fewer than that have an extensive test base. *Reader ability* is the capacity of the individual to make meaning from text. Reader ability, like other latent variables, cannot be directly observed. Rather its existence must be inferred from its effects on measurement outcomes (count correct). A reading test is an instrument designed to detect variation in reading ability. A reading test has three key components:

(1) Text (e.g., a newspaper article on global warming), (2) a response requirement (e.g., answering multiple-choice questions embedded in the passage), and (3) a correspondence table that converts the measurement outcome (counts correct) to a scale value (Lexiles) via a theory (Lexile Framework for Reading). Although the ubiquitous multiple-choice item type dominates as the response requirement in the measurement of reading ability, other task types have found use for specific ranges of the reading ability scale, including retelling, written summaries, short answers, oral reading rate, and cloze. Because the notion of scale unification is still foreign in the human sciences, common practice associates a unique measurement scale with every published instrument. Many dissertations are written that involve development of a new instrument and a new scale purportedly measuring the intended construct in an equal interval metric unique to that instrument. At this writing, the authors are unaware of any dissertation that reports on the unification of multiple measures of the same latent variable (e.g., depression, anxiety, spatial reasoning, mathematical ability) denominated in a common unit of measure. Failure to separate the instrument from the scale has had pernicious effects on human science. Because different reading instruments often employ different task types, different task names (test of reading ability, test of reading comprehension, test of reading achievement), and the aforementioned different scales, it should not surprise us that many test users and reading researchers perceive that these various reading tests measure different latent variables. The unwary are fooled by the fact that looks can be deceiving.

Web-based reading technology (MyReadingWeb) has been developed to accommodate Lexile measurement. Accompanying text and associated machine-generated cloze items is a "key" to score (mark correct or incorrect) reader responses and a correspondence table relating count correct to Lexiles. Any continuous prose can be loaded into MyReadingWeb, and the software will instantly turn text into a reading measurement instrument. The first step involves measuring the text, be it an article, chapter, or book. The Lexile Analyzer computes various statistics on word frequency and sentence length. These statistics are then combined in an equation that returns a Lexile measure for the text. MyReadingWeb uses this measure to "cloze" vocabulary words that are at a comfortable range for a reader's vocabulary who has a reading ability equal to the text readability of the article. A part-of-speech parser then chooses three foils (incorrect answers) that are the same part of speech and have a similar Lexile level as the closed word but that are not synonyms or antonyms of the closed word. Finally, the four choices are randomized before presentation. On average, a response requirement is imposed on the reader every fifty to seventy words of running text.

Counts correct on a text are evaluated by a dichotomous Rasch model in which the location and dispersion parameters of an item difficulty distribution are treated as known. The Lexile measure that maximizes the likelihood of the data is the reader measure reported for each particular encounter between reader and text. A Baysean growth model is used to combine individual article measures within and across days over the complete MyReadingWeb history for a reader.

Table 2.1 presents results from a large study designed to test how well Lexile theory could predict percent correct on machine-generated items like those described above. First-grade through twelfth-grade students (N = 1,743) read a total of 289,345 articles comprised of 194,968,617 words. They spent two years, 157 days, twenty-three hours, sixteen minutes in the program and averaged 150 words read per minute (WPM). Of the 3,051,341 unique cloze items generated by the computer, the participants answered 2,245,741 or 73.90 percent of the items correctly. The model forecasted 2,291,787 correct or 75.11 percent. Figure 3 presents a histogram of differences between theory and observation. One thousand and five (1,005) students had observed counts correct within ± 3 percent of the model expectation for a subsample of 1,325 students.

Table 2.1 Lexile theory prediction on machine-generated items

	Student Count	Mean Reader Measure	SD	Encounters	Words	Time Spent
Overall	1,743	1071L	344	289,345	194,968,617	2y 157d 23h 16m
Grade 1	4	739L	105	47	16,203	2h 36m
Grade 2	217	586L	295	14,449	3,127,596	29d 21h 17m
Grade 3	174	810L	284	22,286	5,644,421	47d 18m
Grade 4	186	946L	295	32,936	10,932,254	74d 19h 24m
Grade 5	164	1074L	221	34,864	15,026,873	92d 10h 12m
Grade 6	175	1130L	180	33,650	16,335,039	80d 20h 48m
Grade 7	171	1171L	229	19,485	9,352,170	43d 1h 36m
Grade 8	164	1281L	252	17,083	8,725,553	39d 19h 37m
Grade 9	149	1285L	254	22,815	19,490,193	81d 19h 21m
Grade 10	130	1268L	229	23,225	21,477,331	89d 11h 49m
Grade 11	102	1324L	151	23,906	26,811,304	104d 19h 16m
Grade 12	107	1353L	157	28,394	35,844,222	128d 3h 13m
Graduated				16,205	22,185,458	75d 17h 49m

WPM	WPM SD	Items	Observed Correct	Expected Correct	Observed Performance	Expected Performance
150	82	3,051,341	2,245,741	2,291,741	73.90%	75.11%
103	44	477	306	308	60.54%	64.53%
72	43	119,307	81,523	85,242	69.02%	71.61%
89	51	193,840	130,819	135,400	68.29%	69.85%
108	58	302,708	202,905	210,746	68.26%	69.62%
121	59	340,605	243,056	251,237	72.35%	73.76%
144	66	338,226	245,232	252,667	73.68%	74.70%
153	70	173,593	127,235	131,027	74.36%	75.48%
158	75	150,612	112,762	114,806	76.09%	76.23%
167	80	264,169	203,037	206,715	77.26%	78.25%
182	80	264,252	199,222	201,473	75.44%	76.24%
191	82	312,976	240,951	246,715	77.38%	78.83%
206	84	384,070	298,797	306,732	78.56%	79.86%
219	86	206,776	159,896	163,092	78.18%	78.87%

Note: Oasis—Reading Data by Cohort—Corinth (MS)
(Data From 2007-06-01 to 2010-04-26)

The best explanation for the close agreement between theoretical comprehension rate and observed comprehension rate is that the Lexile theory and Rasch model are cooperating in providing (1) good text measures for the articles, (2) good reader measure for the students, and (3) well-modeled comprehension rates. The cooperation between substantive theory and the Rasch model evidences cross-sectional developmental consistency (i.e., the theory works throughout the reading range reflected in table 1 [100L to 1500L]). Invoking the "no miracles argument" currently fashionable among philosophers of science of the realist persuasion: the congruence between theory and observation is explained by the fact that the theory is at least approximately right.

In summary, the measurement of human temperature and reading ability if conceptualized in a particular way can be seen to share a common deep structure. Both constructs are latent variables that assign a causal role to an unobservable attribute of persons. Conditioning on the latent variable renders the measurement outcomes (cavity count and count correct) statistically independent. Temperature and reading ability are real entities that can be manipulated, and the effects of these manipulations can be detected. Persons possess a true value on each construct that is approached by repeated measurement but is never precisely determined.

The two attributes apply equally well to between-person variation and within-person variation. For human temperature measurement with NexTemp technology, we can trade off a change in the amount of the soluble additive for a change in temperature to hold the number of cavities that "turn black" constant. Similarly, for reading, we can trade off a difference in text readability to hold constant the count correct. In both cases, enough is known about the measurement procedure and the relevant active processes that two persons with equal temperature or reading abilities can be made to produce different measurement outcomes (cavity count or count correct) by systematically manipulating the respective instruments. In short, the two latent variables are under precise experimental control, and that is why an indefinitely large number of parallel instruments can be manufactured for each construct.

Both thermodynamics theory and Lexile reading theory force a distinction not made before the theories were put forth. The sensation of hot and cold was formalized as temperature and was later distinguished from the common parlance synonym "heat." Reading ability was likewise distinguished from the common parlance synonym "reading comprehension." The former is a text independent characterization of reader performance, whereas the latter is a text dependent characterization. Both theories have made extensive use of the ensemble interpretation first proposed by Einstein (1902) and Gibbs (1902).

Oasis Reading Data (Grades 2–12): Observed
Minus Expected Performance
Aggregated by Student (n = 1,325)
9/15/07–5/27/08
Filters: WPM, Two Hundred Items

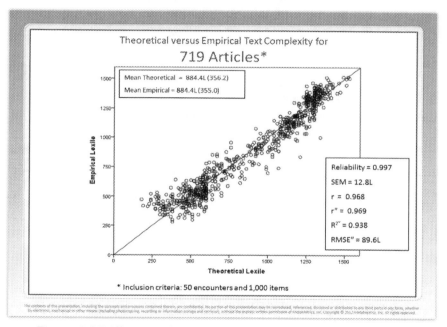

Figure 2.3 Differences between Lexile theory and observations

The two constructs, temperature and reader ability, figure in laws that are strikingly parallel in conception and structure. The combined gas law specifies the relationship between volume and temperature conditioning on pressure. Specifically, log pressure and log volume — log temperature = a constant, given a frame of reference specified by the number of molecules. Similarly, the reading law specifies the relationship between reader and text conditioning on comprehension rate. Logit transformed comprehension rate plus text measure — reader measure = the constant 1.1, given a frame of reference that specifies 75 percent comprehension whenever text measure = reader measure. Therefore $a + b - c$ = constant, holds for both the combined gas law and the Lexile reading law (Burdick, Stone, and Stenner 2005).

Finally, the NexTemp and MyReadingWeb technologies share several additional features: (1) they are both inexpensive (NexTemp cost is nine cents, MyReadingWeb's cost per item is fractions of a penny), (2) they function within an intended range and are useless outside that range, (3) the respective technologies produce instruments that are one-off and disposable, (4) both instruments are theoretically calibrated, and this produces generally objective measures, (5) both are readable technologies—the user does not need to understand even the rudiments of thermodynamic theory or Lexile theory to produce valid and useful measures, and (6) both can measure growth or change within and between persons.

In conclusion, although closed knowledge systems have not been developed for health outcomes, corollaries between temperature and reading ability presented in this paper should offer valuable insights to their formulation. For example, functional assessment in rehabilitation currently defined by separate CAT measures could be consolidated into a closed system defined by deep structure common across functional measures. Separate functional items could be reduced to a single structure that includes not only outcome but intervention constructs. In this system, variation of intervention treatment values would be systematically related to outcome measures. Consequently, treatment effectiveness could be specified in advance of delivery and terminated at maximum effectiveness.

Box 3: New Bottles for Old

Drawing inspiration from Aesop's phrase "What pleasures cling round the old instrument," we find utility in putting old words in a new context to usher in ideas. We begin with this paradigm:

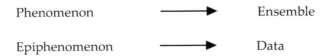

Phenomenon   ⟶   Ensemble

Epiphenomenon   ⟶   Data

During the past century of measurement science, great attention has been given to data and in particular to items. This has resulted in a literature that is rich in methods for item construction and item analysis.

We suggest realigning the focus from the data/items to the ensemble or aggregate as a useful way of moving ahead (Brody 1993).

In effect, it is the mechanism/instrument that should be focused upon, lest we miss the forest for trees. Data is relevant and important, but it is the mechanism by which we make measures that should receive focal attention. Theory is based upon relationships, not upon isolated facts. If you want to check the temperature, it is the instrument that supplies the result. It is not the numerals or any singular aspect of the process. The mechanism taken as a whole supplies an answer. The bathroom scale is the mechanism used for producing "weight" in specified units. Both mechanisms produce outcomes irrespective of time, place, or person. Mechanism is not solely mechanical; rather, it is the totality of the structure for providing an answer to questions of temperature or weight.

Phenomenon/mechanism in this discussion represents the collective, the totality, the aggregate, and the outcome when employing the ensemble. Data constitutes an epiphenomenon. From the consequence of theory and data emerges a mechanism, with data no longer the focus. Spending one's attention solely on bits of data will not produce a cohesive whole. The thermometer and scale are composite mechanisms. They consist of an assemblage of parts and pieces integrated by cohesive ideas that ultimately lead to realization by a theory for measuring: one for temperature, and another for weight.

Success is achieved when a theory can be brought into realization by a mechanism that will substantiate the desired effect. It may take time, and it may require traveling false paths, but if the goal is well conceived and the theory is sound, getting there can be mechanized in some fashion. We move substantively forward from an assemblage of bits and pieces to an integrated whole.

The ensemble becomes that whole in this discussion. The ensemble represents a great leap beyond data. Measuring constructs by a mechanism evolves along a common historical course. Early in construct history, the first measures are subjective, awkward to implement, inaccurate and poorly understood—the king's foot as a measure of length is one illustration. With time, standards are progressively introduced, common metrics are imposed, artifacts are adopted (e.g., the meter bar), precision is increased, and use becomes portable and ubiquitous. Finally, the process of abstraction produces another leap forward, and the concrete artifact-based framework is left behind in favor of a theoretical process for defining and maintaining a unit of length (e.g., oscillations of a cesium atom for length of time). Human science instrumentation evolves along a pathway of increasing abstraction. In the early stages of development, a construct and unit of measurement are inseparable, stemming from a single arbitrary source. Over time, multiple instruments come to share a common metric, and finally the construct is explicitly specified by an equation. When a specification equation exists for a construct and accounts for a high percentage of the reliable variance, the ensemble and the construct are no longer operationalized by a bank of data but rather by the causal recipe for generating computer-realized items with prespecified attributes. The items do not drive the process. It is the mechanism (theory to construct) that utilizes items as necessary, making items no longer sacrosanct. They serve but do not dictate or direct operations.

When we achieve multiple mechanisms for operationalizing the process, we leap forward once more by the capacity to provide predictive control over the process. Predictive control is the hallmark of success. Our efforts are substantiated (albeit probabilistically), or else we do not know what we are doing. Quality control must be continually applied because no mechanism operates successfully without continual monitoring. However, in this instance, we are monitoring a known theory, process, mechanism, and outcome. Any measurement embodies a construct theory, which can be thought of as a story about what it means to move up and down a scale (Stenner, Smith, and Burdick 1983).

It is common sense that causal relationships are those utilized for manipulation and control. If C is thought to cause E, then it should be possible to manipulate C to produce changes in E. Such theories concerning causation can be found in philosophical works by Gasking (1955), Collingwood (1940), von Wright (1971), and Menzies and Price (1993). Early applications were made by Haavelmo (1944) to econometrics. Cook and Campbell (1979) provided paradigms for making evaluations in the field illustrated by their comment, *"The paradigmatic assertion in causal relationships is that manipulation of a cause will result in the manipulation of an effect* ... Causation implies that by varying one factor I can make another vary"* (36, emphasis in original).

A similar approach can be found in the literature on structural equations or causal modeling (Hausman 1998; Pearl [2000] 2009; Rubin 1986; Sosa and Tooley 1993; Woodward 1984, 2011). Pearl ([2000] 2009), for example, straightforwardly defines the "causal effect" of X on Y as associated with the "realization" of a particular value x of X influencing y. An elementary linear regression equation may also express a casual inference. The problem rests with substantiating the mechanism.

Scales and thermometers are "causal" when viewed as mechanisms for fabricating outcomes engineered from the environment and the thermometer, or the scale and a person. Both mechanisms "cause" outcomes and meet the essential characteristics of a measure. They are general, utilitarian, portable, and replicable.

Taking a new look at the *mechanism of measuring* from the point of view of the ensemble and not from the pieces, from the point of view of causal process and not a description of data, suggests a richer future.

Box 4: Invariance

Invariance typifies a property that remains unchanged under a transformation. In general, scale invariance implies that the characteristics of laws or objects do not change when multiplied by a common factor, technically called "dilatation." Scale invariance is generally a straightforward mathematical problem, but when considered in conjunction with a theory or in relation to a substantive example, it can prove more problematic. Invariance does not imply overall stasis in a system but rather no change under a current operation. Invariance with respect to data procurement procedures belongs in this domain. Invariance itself is undefined until placed in some context. Invariance depends upon its specific application and should be interpreted in that context only.

There are also studies of multiple invariance, demonstrated by Lyle Jones (1958), who produced a summary of invariances demonstrated for results using the method of successive intervals.

1. The number of categories in a rating form, and the anchoring phrases associated with the categories—identity transformations
2. Assumption of normal subjective distribution replaced by assumption of normality of errors with repeated administrations—linear transformation
3. Samples of subjects and classes of behavior regarding the same stimuli—linear transformation
4. Samples of subjects and sets of stimuli from the same general class—invariance of category widths to an identity transformation

Stevens (1951) states,

> The scientist is usually looking for invariance whether he knows it or not. Whenever he discovers a functional relationship his next question follows naturally: under what condition does it hold? ... The quest for invariant relations is essentially the aspiration toward generality,

and in psychology, as in physics, the principles that have wide application are those we prize (20).

Each of the four classes of scales [nominal, ordinal, interval, and ratio] is best characterized by its range of invariance – by the kind of transformations that leave the 'structure' of the scale undistorted. And the nature of invariance sets limits to the kinds of statistical manipulations that can be legitimately applied to the scaled data. (21)

Achieving invariance leads to generality and to predictive success. Invariance is required for validity to be realized.

Box 5: Counting and Measuring Revisited

A renewed examination of the differences between counting and measuring may be warranted. Psychometric dogma suggests that measuring trumps counting, without reexamining the evidence for making such a statement. Another cliché suggests that the quantitative swamps all other considerations in metrology. Counting and measuring are inextricably intermixed with such matters as qualitative/quantitative, transformation, analogy, symmetry, and unit. At this time, we can only reexamine counting and measuring, and only just in an introductory manner.

Consider the lowly hourglass. Sand is slowly passing through a narrow passage to mark out time in an analogous process. Up-ending the hourglass is equivalent to counting the hour(s) or minutes. Might some user have made a mark at the halfway point, the quarter, or even less? Could somebody have attempted to count the grains in search of an absolute subdivision?

We clearly have a measuring analogy whereby time (more properly, duration) is marked out by the passage of grains of sand from the upper glass to the lower one. While the hourglass is calibrated to the hour [recall the three-minute glass for eggs] we might speculate on what the early ponderings about the subdivisions mentioned above indicated to thinkers of the past contemplating the counting or measuring of time. What about the grains of sand? What might their early musings imply about counts and about measures? Might we have almost come close to having grains of sand instead of seconds to compose the hour? Archimedes utilized a number on the order of 10^{52} grains of sand to represent the size of the celestial universe. How do we denominate more modestly what is conjured up from this glass of sand? Is the hour, a subdivision, or grain to be considered a count or a measure? Is it a unit?

Must all measures begin with a crude count? Is that because what was first counted was an object or a substance? The inch and foot were early and crude proxies, but they have survived. Can we not count abstractions also? If so, does this not also imply that abstractions might be contaminated by mixing them with delusions, hallucinations, and mirages? Furthermore, is our count or measure to be considered qualitative or quantitative? If we cannot answer these questions precisely for an hourglass, how do we expect to answer validly when addressing more complicated dimensions? We could, of course, dismiss all this inquiry as unnecessary and irrelevant, but is that valid, or is it only to avoid facing fundamental but difficult questions?

Is a count of the natural numbers to be deemed continuous or qualitative? Are n + 1, n + 2, and so forth, qualitatively different from *n* by their expression as different numerals? Are they not also isolated segments of a continuing series? Is not matching the set of numerals to a set of objects or substances a case of measure? If so, the measure of the set was also the count. Does this hold when applied to the process of selecting apples? Do we measure or count a selection of Granny Smith apples? Does the process change when an abstract variable is considered rather than apples?

What is required is to evaluate the properties of counting and measuring as contemplated, postulated, and applied. We cannot deal with the words stripped of their context. Symmetry between counting and measuring needs further examination, as do the words *qualitative* and *continuous*, together with all the other words suggested earlier. Rasch measurement requires some rethinking regarding uses of these words.

Luce (1996, 78–79) discussing the dialogue between empirical science and measurement writes:

> Classical measurement theory is characterized primarily by topics in physical measurement leading to representations [...] designated representational measurement.

Luce went on to suggest three overlapping stages in this process: classical, contemporary, and a third stage that constitutes an "attempt to put our knowledge about measurement to use in devising theory."

Narens and Luce (1986) enumerated fifteen problems with representational measurement theory (RPM) suggesting all is not right with this approach. While the representational paradigm may prove helpful, it is best interpreted as a "consequence" having originated from imputed standards such as the "royal foot," a substantive physical object. Further application of this approach produced variables advanced by theory and experiment such as temperature.

RPM might be nomenclature for hypothesized super-stantiated variables. A super-stantiated variable is proxied as if possessing physical qualities (Hans Vaihinger 1925). "As if" implies a fiction of substance but without grounding as such. There is nothing wrong with this approach as long as one does not assume more than is warranted when applying this fictive strategy. The legal and judiciary fields have largely operated in this mode.

The earliest stages of measuring might be advanced as "fundamental," except that the word already appears in too many different contexts to be useful. However, this investigation borrows its impetus from a well-known philosopher's work (Heidegger), who sought grounding to all metaphysical speculations. These remarks concern finding the way back into the ground of metrology. A bit more attention to the grounding elements might offer some insight into today's measurement controversies and redirect investigations.

Dantzig (1954) reminds us that a cardinal number implies matching, not counting. This reminder can be applied to the concept of *number sense* defined as characteristic of a class of similar collections. Dantzig (1954, 6) further describes the number sense of the Thimshian of British Columbia by reporting the observations of the anthropologist Boas, who noted "seven distinct sets of number words for flat objects, animals, round objects, time, men, long objects, and canoes" in their culture, and thus required different number sets for different substances. In this setting, the numeral seven depends upon its application. For Danzig, substance precedes naming and counting, which occurs later in human development. Dantzig argues that the concrete precedes the abstract, noting that without the need for counting, we immediately grasp "one-ness, two-ness, and three-ness" but do not easily grasp the size of larger collections. These elementary whole numbers function like shorthand for the matching process. We grasp them immediately as in selecting apples when making a purchase. Larger collections require introducing the process of order followed by counting. Some early societies (Dantzig 1954) counted "one, two, three … more." Word labels describe these aggregated collections directly, but counting enabled further development in enumeration. The concept of an abstract number represents a great leap forward from these modest beginnings grounded in physical objects—substance. Order has always been fundamental to mathematics. Integers, rationals and reals, all possess an order of magnitude. Points on a line, dimensions, and the concept of limits likewise may follow a serial conception of order.

Astronomy is one of the earliest of the sciences. It was not grounded in experimentation but in observation of the heavens, planets, and particularly the stars, for which two characteristics have long been prominent through the ages. Since early times, stars have been observed to vary by color from blue-white to amounts of redness. Size is a second early observation by which visual stars were ordered. Today, astronomers differentiate three magnitudes: apparent, absolute, and bolometric, but in early times, it was simply apparent brightness.

Plotting size by color against apparent brightness produces what is known as a Hertzsprung-Russell diagram (Baker 1978). Today, Hertzsprung-Russell plots of brightness or luminosity against magnitude together with modern spectography have produced derived estimates of mass, luminosity, distance, heat, and so on, all made from great distances. It is not these later estimates that are our concern but how important simple observation together with the elementary grounding of *comparison and order* to substances greatly contributed to later stages of insight and knowledge.

Early measures focused upon elementary observations that ground comparison and order prior to the development of counting. This is seen in the early observation and comparison of stars by their visual order in brightness and size. These two subtle attributes—comparison and order—mark the grounding of early measures. The grounding of measures may include the small group-sense of collections and classes formed by comparing and by ordering. Such collections might later be enhanced by counting and the concept of abstract number.

Elementary science was grounded by the observation of substance/objects leading from speculation to hypotheses and knowledge. This approach dramatizes the dialogue between empirical science and theory—from simple observation to more sophisticated insight, inferences, and applications. Implicit in this process, but extremely important, is the fact that these elementary processes were driven by substance/object observations and the elementary processes of comparison and order. This development leads to abstraction, not the other way around. Substance/object observations pushed measuring objectives into being via comparison and order. Measures were not created a priori and then applied. Nor are they even required a priori, as indicated by Guttman (1971) and Rasch (1977).

Consider once again selecting apples. Our preference is Granny Smith, especially for its firm, tart qualities. Green is their color, but not all green apples constitute a Granny Smith. What we choose not to address is equally important to what we focus upon. Size is relevant because a Granny Smith is not as small as a grape or as large as a watermelon. When selecting them in a store, we focus upon a Granny Smith apple and ignore bananas, nuts, and all other produce equally tasty or preferred by others. Were we permitted to sample them, our choice might be perfected, but that is not proper. If a handful suffices, more or less, no counting occurs. Comparisons have been sufficient; otherwise we count or weigh, which introduces advanced measurement not required to make such an elementary selection. There is comparison and order grounding our selection. We began with a substance/object and personal observation guiding our choice of Granny Smith above all other apples, even other green ones, by virtue of the tart taste we prefer.

We began this elementary process of selection by comparing through qualitative decision making of substances/objects. From qualitative decision-making we proceed to employ simple comparisons and elementary ordering. Our math involves matching at best, and counting is not usually required. However, from these elementary processes, it is possible to increase knowledge by improving decision-making with more advanced measuring methods, such as counting and weighing.

Many day-to-day decisions are accomplished by observation guided by attention to a substance or object, together with simple comparing and ordering. These elementary processes ground the measuring process. Comparison is employed not only for practical tasks but also as the grounding for science. Order quickly springs forth as an important tool (Stone and Stenner 2014).

Rasch (1977) appears to have been pursuing this approach when he specified that measurement begins with qualitative discernments made from systematic comparisons. He called such comparisons "indispensible in scientific statements." Rasch (1977) draws attention to their ubiquity by means of observation:

> … comparisons form an essential part of our recognition of our surroundings: we are ceaselessly faced with different possibilities for action, among which we have to choose

just one, a choice that requires that we compare them. This holds both in everyday life and in scientific studies. (68)

The making of comparisons is followed by their order as the second of the two elementary processes grounding the measuring process, in the same way that early observers first compared and ordered the stars by color and apparent brightness. It is these elementary processes of comparing and ordering that ground initial observations. Without achieving success in these elementary processes, no progression to measuring is possible. We argue that such elementary processes ground whatever follows, and everything else builds upon these elementary processes. How we compare and order grounds further development in measuring. These elementary strategies ought not be minimized by esoteric discussions and studies that ignore the elementary grounding that undergirds all measuring.

Observations of objects and the elementary processes of comparing and ordering are indispensable to science. The dialogue of empirical science and theory must rest upon grounding in comparison and order, but it is important to remember that questions about substance and objects were the origins for measuring.

Box 6: Parameter Separation

Separation of the item and person parameters is an important and useful consequence of Rasch measurement. Rasch argues that "objectivity" is achieved when a comparison of any two objects is "independent of everything else within the frame of reference other than the two objects which are to be compared and their observed reactions" (Rasch 1997, 77). Therefore, when the person or item parameter is isolated (separation), only the two items or the two persons remain in the equation. One intriguing aspect of the parameter separation issue is that Rasch himself did not fully appreciate its importance until Ragnar Frisch, the Norwegian economist, called attention to it several times when Rasch was explaining his work to Frisch.

Rasch's argument can be presented in the following steps:

Step 1 (basic formula):

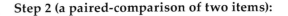

$$\text{probability } [\, x_{ni} = 1 \,] = \frac{e^{(\alpha v - \delta i)}}{1 + e^{(\alpha v - \delta i)}}$$

where
x_{ni} = a correct response by person n
α_n = the ability parameter of person n
δ_i = the difficulty parameter of item I

Step 2 (a paired-comparison of two items):

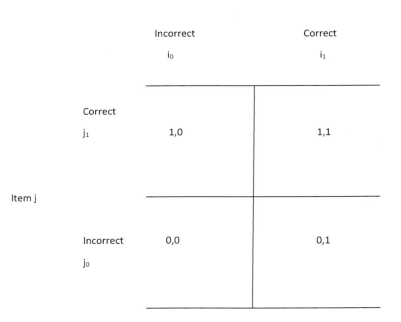

Only cells 1,0 and 0,1 provide information, i.e. one item correct and the other incorrect. Items 1,1 and 0,0 provide no information for comparison because both are either correct or incorrect and no information about their relative difficulty is available.

Step 3 (the logistic expression):

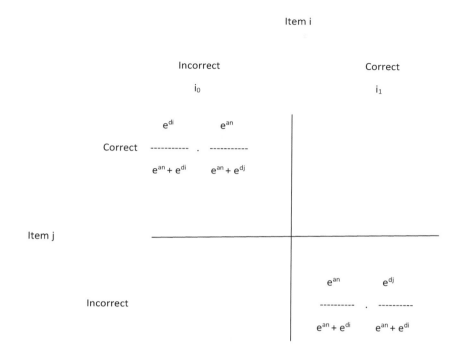

Step 4 (the conditional probabilities):

The conditional probabilities:

Probability $[a_{nj} = 1 \mid a_{ni} + a_{nj} = 1]$

$= e^{(an + di)} / [(e^{an + di}) (e^{an + dj})] / e^{an} (e^{di} + e^{dj}) / [(e^{an + di}) (e^{an + dj})]$

$= e^{dj} / (e^{di} + e^{dj})$ [2]

so that probability $[a_{nj} = 1 \mid a_{ni} + a_{nj} = 1] = e^{di} / (e^{di} + e^{dj})$ [3]

    The ability parameter a_n has been eliminated from equations 2 and 3 leaving only the two item difficulties. What becomes especially useful is that by making multiple pair-wise comparisons of this kind, it is possible to estimate the relative locations of any items on the measurement variable.
    When test data conforms to a Rasch model, "objective measurement" is

possible because the comparison of two item parameters does not depend on the details of the persons who took the items. Likewise, the comparison of two person parameters does not depend on the details of the items used to make that comparison. Aggregating all the multiple paired-comparisons allows the locations of items and persons to be located on the same variable.

Estimation in Practice

The probabilities of a correct responses to item i and j given that one is correct and the other incorrect can be estimated by observing the respective counts of those persons attempting both items. For every pair of items in the test, we accumulate the observed b_{ij} and b_{ji} to estimate the relative difficulty of the two items under examination. Rasch ([1960] 1980, 1966) described this approach and Choppin (1965) developed a program for pair analysis. The more common approach is a UCON method adopted in most Rasch estimation programs. However, the pair analysis, illustrated here, is especially helpful in showing Rasch's parameter separation. WINSTEPS (Linacre 2002) and FACETS (Linacre 2002) are well known and commonly used in the states. RUMM (Andrich, D., et al., 2000) is also well known and utilized.

3

MAPPING VARIABLES

"Map of the Variable" might be considered the beginning and end of assessment. However, we must immediately add that variable construction and mapping are never complete; the processes are always ongoing. Variables require continuous attention to both their development and maintenance. The map of the variable is a visual representation of the history and status of the variable construction. It is a pictorial representation of "the state of the art" in the construction of a variable

The Origins of Mapping

Maps are visual guides that ground us in a stable frame of reference and provide a sense to direction. Expressions implying vision such as "Do you see?" and, "Show me what you mean," and, "Now I see," are frequently used. These expressions testify to the visual power inherent in pictorial representations and further conveyed in speech and writing. Maps visualize the scope of our knowledge. They are indispensable to planning and travel. Mapmaking has great utility.

The inability to understand or make use of maps is a handicap to understanding the world. The earliest maps used naturally occurring phenomena—celestial and terrestrial—to identify features.

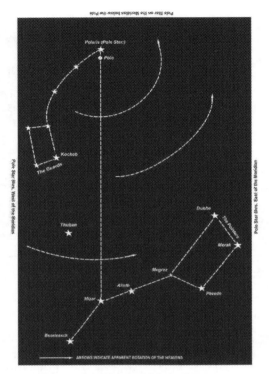

Figure 3.1 Big and Little Dippers
From Lecky, T. S. 1910. *Wrinkles in Practical Navigation.* London.

Figure 3.1 is a map showing the Big and Little Dippers. We use the "pointers" of the Big Dipper to locate Polaris, the pole star. Although both dippers move, they rotate around Polaris, which appears fixed. From Polaris we determine north. More comprehensive maps include the popular figures of the Zodiac, other constellations, and lesser-known celestial features. These maps provide visual pictures of the evening sky throughout the year.

Terrestrial markers also serve as guideposts. A lake, a river, or a mountain may be used to orient and anchor locations. Celestial and terrestrial maps have been used for centuries and are occasionally brought into relationship with one another.

The map of Teotihuacan reveals an urban grid as deliberate as Pierre L'Enfant's plan for Washington, DC. The grid used two principal, almost perpendicular alignments. The east-west axis led from a spot near the Pyramid of the Sun to a point of great significance on the western horizon. Stuart (1995, 15) reports astronomer-anthropologist Anthony Aveni of

Colgate University explanation that on the day the sun passes directly overhead in the spring, about May 18, the revered Pleiades star cluster makes its first annual predawn appearance. It was at this point on the western horizon that the Pleiades set. A second theory notes that the sun also sets here on August 12, the anniversary of the beginning of the last great Mesoamerican calendar cycle. Many scholars believe this occurrence to have begun August 12, 3114 BCE. Whatever the astronomical motive for the axis, it was considered so important that the channel of the San Juan River, which crossed the center of the site, was rerouted to align with it.

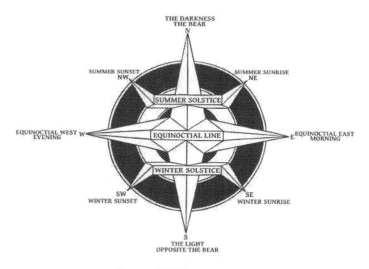

Figure 3.2 Compass rose
From Brown, L. A. 1949. *The Story of Maps*, 123. New York: Dover.

Figure 3.2 shows the compass rose and the directions derived from making observations about the rising and setting of the sun throughout the year. This diagram of a naturally occurring event illustrates how mapping natural phenomena leads to abstractions. A simple east-west orientation observed from the rising and setting sun also gave impetus to its perpendicular north-south. From this four-point orientation evolved the eight- and sixteen-point divisions of the compass rose.

A map with the compass rose serves as an analogy, an idea that pictures an abstraction. While a map may initially seem simple, superficial, incomplete, and perhaps even inaccurate, it still may be useful for various purposes. A map shows the current status of what is known about a domain. The compass rose provides an orientation. It offers a perspective and means for interpretation and application. It moves understanding from simple portrayal to generalization, from particular to general, and from representation to abstraction.

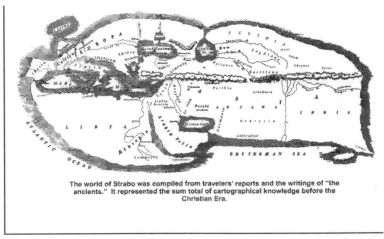

The world of Strabo was compiled from travelers' reports and the writings of "the ancients." It represented the sum total of cartographical knowledge before the Christian Era.

Figure 3.3 Mediterranean Sea area by Strabo

From Bunbury, E. H. 1879. *A History of Ancient Geography*, Vol. II, Pl. III. London.

Figure 3.3 is a map of the known world at the time of Strabo (63 BCE–24 CE). It shows the lands around the Mediterranean Sea and points eastward. By current standards, it is crude and inaccurate, yet we can observe its general validity. This observation illustrates another important aspect of mapping. Maps, by their very nature, invite improvement. Every edition of a map draws attention to its accuracy and to its inaccuracy. Each new edition incorporates changes, resulting in a more accurate version than the previous one provided.

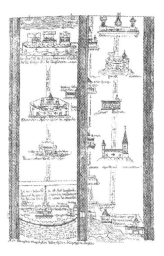

Figure 3.4 Pilgrims' guide, London to Jerusalem
From Jomard, E. F. *Les Monuments de la geographie, Paris, 1842–1862*, Pl. V(2).

The map in figure 3.4 guided pilgrims on their journey from London to Jerusalem. An intriguing feature of this map is its linearity marked out by pictorial highlights along the side indicating "equally spaced" sites along the way. This example illustrates another aspect of mapping, highlighting important locations. When maps display highlights, they enhance understanding by offering a perspective for interpretation and use. The day-to-day sites for pilgrims provided refuge. Spaced by day, the pilgrims could expect regularity in their journey. Employment of a day's journey offered a practical way to visualize the journey ahead and plan accordingly.

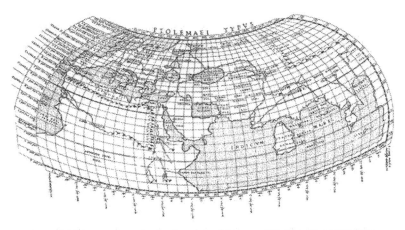

The world of Ptolemy according to a Venetian editor, 1561. Longitude is expressed in fractions of hours east of the Fortunate Islands while latitudes are designated by the number of hours in the longest day of the year.

Figure 3.5 Lines of longitude and latitude
From Ruscelli, G. *La Geografia di Claudio Tolomeo*. Venice: 1561.

Figure 3.5 superimposes upon an outline of the known world of 1561 the lines of longitude and latitude. This addition illustrates the benefits of introducing an abstraction, in this case longitude and latitude, upon the natural contours of land and sea. Abstractions enhance maps by advancing generalization. Such an abstraction moves the map to every increasing application.

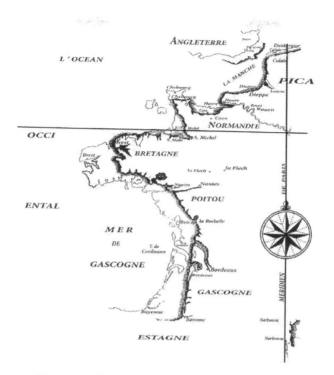

Figure 3.6 Western France in Napoleon's time
From *Memoires de l'Academie Royale des Sciences, Paris, 1729*, Tome VII, Pt. I.

Figure 3.6 shows Napoleon's perspective of Western France. The corrections to an earlier map given in this new edition are noted. The corrections are greatest where Western France is adjacent to the Atlantic and England. This detail of the map could suggest that perhaps the creation of the map was prompted by the idea of a possible invasion. It is clear that when more natural reference points are available, error occurs less often. When the natural markers are wider apart, there is a greater possibility of error between markers. Interpolation and extrapolation are weakest when the markers are few and widely spaced. They are at their weakest in this condition because we cannot assume what occurs between them and beyond.

These early maps illustrate several important points:

1. Maps are useful pictures of experience. Our first and quickest form of orientation remains visual.
2. Inaccuracies in details and features of maps require successive and inevitable corrections.

3. Useful abstractions, such as longitude and latitude and external markers, enhance mapping and improve knowledge.
4. Increased knowledge produces greater accuracy. The process is cyclic and continuous.

Maps of Variables

Since a map can be an abstraction of a variable, map topography is applicable to psychometrics. The variable implied by a test can be pictured as a line. A line with direction marked by an arrow is the beginning of a map. Continuous improvement is irresistible. Corrections are invited. The more information we gather about the variable, the more accurate our representation becomes. Finally, this pictorial representation of the variable invites further abstractions that generalize understanding. Rudolph Carnap (1966) wrote:

> The nineteenth-century model was not a model in this abstract sense (i.e., a mathematical model). It was intended to be a spatial model of a spatial model structure, in the same way that a model ship or airplane represents an actual ship or plane. Of course, the chemist does not think that molecules are made up of little colored balls held together by wires; there are many features of his model that are not to be taken literally. But, in general spatial configuration, it is regarded as a correct picture of the spatial configuration of the atoms of the actual molecule. As has been shown, there are good reasons sometimes for taking such a model literally -- a model of the solar system, for example, or of a crystal or molecule. Even when there are no grounds for such an interpretation, visual models can be extremely useful.

> The mind works intuitively, and it is often helpful for a scientist to think with the aid of visual pictures. At the same time, there must always be an awareness of a model's limitations. The building of a neat visual model is no guarantee of a theory's soundness, nor is the lack of a visual model an adequate reason to reject a theory. (176)

There are three important uses of maps:

To direct ... where we are planning to go;
To locate ... where we are, along the way; and
To record ... where we have been.

A map implies the beginning of test construction. In the beginning, a map defines first our intentions. At the end, it is a realization of progress to date. Maps of variables are never finished. They invite constant correction and improvement. Maps embody abstractions derived from experience. Maps connect the world of the mind to the world of experience. Abstraction is validated by correspondence to experience inasmuch as experience is best understood by abstraction. Mapping promotes the dialogue that must take place between these two worlds in order to communicate constructively. A map produces a visual, operational definition of a variable. While maps are necessarily only models, their pictorial representation invites continual correction producing ever-increasing accuracy (Stone 1995b).

Graphs as Maps

Graphs of functions are maps showing the relationship between two variables. Graphs make it easy to tell simply by looking whether a useful function is emerging. Graphs make functions recognizable and familiar. As discussed prior, a line or arrow can be used to orientate us to a map. The familiar curves for a parabola, a quadratic relation, and functions like the undulating sine curve are easily recognized by their shapes. The graphs of functions are maps as familiar to their users as road maps are to motorists. They aid understanding by simplifying the process and allowing us to "see" a complex representation. According to Danzig (1954):

> Descartes gave the world the *analytical diagram* which gives at a glance a *graphical* picture of the law governing a phenomenon, or of the *correlation* which exists between dependent events, or of the *changes* which a situation undergoes in the course of time. (192, original italics)

A Map as an Analogy

Measurement is made by analogy. The most efficient and utilitarian measures rely upon visual representation. The ruler, the watch face, the mercury column, and the dial are common analogies used to record length, time, temperature, and weight. The utilitarian success of analogy in these measuring tools is demonstrated by their ubiquity.

Oppenheimer (1956, 129) said of analogy, "It is an instrument of science." He went on to say, "Analogy is inevitable in human thought." By these comments and others, he explained the role of analogy in facilitating thought and discovery. It was not, he said, to be the criterion of truth.

The implications for psychometrics from this survey of maps suggest the following points:

1. The "intended map" of the variable is the idea, plan, and best formulation of our intentions.
2. The "realized map" of the variable derived from item calibrations, person measures, and standard errors implements the plan.
3. The continuous dialogue between intention (idea) and realization (data) produces and maintains the validity of the variable.
4. A map of a variable is often the scope and sequence of instruction or learning. It shows how to arrange instruction and guide relevant assessment.
5. The progress from instruction and resultant learning can be located on the variable. Growth can easily be seen and measured.

There are shortcomings to maps, and this is especially evident if their use is pushed to extremes. What Kaplan (1964) presents in his discussion of the shortcomings of models also applies to maps. We must be careful not to expect too much of a map and ascribe more substance to what is produced than can be justified. Constant monitoring of map building is necessary. Monmonier's (1996) critique *How to Lie with Maps* presents in a useful and amusing way the fallacies that can result from viewing a map as a "finished product" rather than as a "fiction," an approximation of the outcome and one that is in a process never to be completed. Braithwaite ([1953] 1956) has cautioned, "The price of the employment of models [maps] is eternal vigilance."

Sometimes maps serve distorted and humorous views. Many will be familiar with the "New Yorker's Idea of the United States" promulgating the city and subverting everything west to minimal representation. There is an earlier and similar version with Boston featured as prominent. In England, the Doncaster and District Development Council (Gould and White 1974) published "Ye Newe Map of Britain" depicting land north of London. Its goal was to move its citizens' vision of opportunity beyond their narrow local horizon by a subtle ploy. The map depicts civilization's end at Potter's Bar on the outskirts of the city. Scotland is just north of the

Arctic Circle pictured in the land of icebergs, dog sledges, and close to the North Pole. The effect of this ploy is not known.

Beyond these humorous examples, Ekman and Bratfisch (1965) and Dornič (1967) began a series of studies with Swedes investigating how their local emotional involvement corresponds to their knowledge and awareness of happenings in other geographical locations in Sweden. They concluded that the degree of external interest corresponds to the square root of the distance. From these studies, Gould and White (1974) further postulated an "information surface" to describe a mental variable consisting of concentric, isobar-like patterns constructed from determining the degree of knowledge beyond one's immediate surroundings. On these maps, the portrayal of information possessed about other geographical areas appears to be a function of the local population size and distance:

Information – F (population, distance).
Another similar study from Gould and White (1974) using Illinois students suggested
Log Information = -1.38 + 0.87 Log Population – 0.40 Log Distance
as a model equation.

These interesting studies along with the creative employment of pictorial and graphic presentations, such as the ones given by Tufte in *Visual Explanations* (1997) and other publications, indicate the rich and productive field that enlightens us by the construction of ingenious mental maps such as the ones depicted by Gould and White (1974). Hence, maps have morphed from terrestrial, celestial, and human-referenced physical domains to cognitive fields bringing the promise of more interesting discoveries and results.

Maps in Psychometrics

Psychometric maps serve as the plan for instrument development and revision. The map of a variable is a blueprint for a test. When a map is logical and well constructed, its implementation can be a straightforward representation in the form of ordered items and persons. When the map is empirically verified, it documents the successful realization of an idea. The map of the variable depicts both the idea and its realization in the form of calibrated items and measured persons. Binet's work in test development implies mapping, although he provided no map. Binet (1916) describes his test:

First of all, it will be noticed that our tests are well arranged in a real order of increasing difficulty. It is as the result of many trials, that we have established this order; we have by no means imagined that which we present. If we had left the field clear to our conjectures, we should certainly not have admitted that it required the space of time comprised between four and seven years, for a child to learn to repeat 5 figures in place of 3. Likewise we should never have believed that it is only at ten years that the majority of children are able to repeat the names of the months in correct order without forgetting any; or that it is only at ten years that a child recognizes all the pieces of our money. (185)

Binet describes how data from experience established a hierarchy of item difficulty. He relied on "numerous" replications of ordered items to produce a sequential order with the level of accuracy he desired. One might say, as Binet did, "It matters very little what the tests are so long as they are numerous" (Binet 1916, 329).

Binet focused on two essential psychometric points:
1. Item arrangement by difficulty order
2. Numerous items, sufficient for precision

Note Binet's designation of items as "tests," which is exactly what each item truly is. Those familiar with Binet instruments recognize that each "test" presented a problem to be solved, each one presenting a different challenge. He was correct in designating "test" for what we call "item" inasmuch as test more accurately describes the puzzle, problem, or question he constructed. What we designate "test" as but an ordering of "tests" and more appropriately a scale of "tests" better labeled a scale or instrument.

There is no other way to build an instrument, except to do as Binet did: begin with an idea for a variable, illustrate the variable by tests [items], order them by their intended and verified difficulty, and measure persons by their locations among the items solved successfully and those not solved successfully as spaced along the variable defined by the hierarchy of "tests" [items]. The hallmark of Binet's early efforts is the benchmarking of items and persons on a variable. He clearly envisioned a mental map of what he intended to produce. Thurstone's *An Absolute Scale of Binet Test Questions* (1927) follows as figure 7 [Thurstone's figure 6].

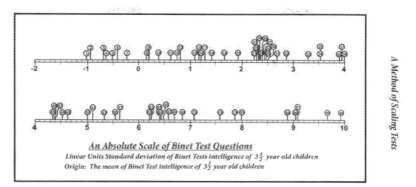

An Absolute Scale of Binet Test Questions
Linear Units Standard deviation of Binet Tests intelligence of $3\frac{1}{2}$ year old children
Origin: The mean of Binet Test intelligence of $3\frac{1}{2}$ year old children

Figure 3.7 Thurstone's *An Absolute Scale of Binet Test Questions*
From Thurstone, L. L. 1925. "A method of scaling psychological and Educational tests." *Journal of Educational Psychology* 16 (7): 433–448.

Thurstone's map provides insight into a sample hierarchy of the Binet "test" sequence for age three-and-a-half-year-old children using data from Burt (1922, 132) for three thousand London children. It is not just a scale of "order" that Thurstone established for each Binet "test" but a measure of the "distance" that exists between individual tests. From this map, we can evaluate the "absolute" difference between individual tests, and the ratio of distance between any two or more tests we seek to compare. Only one map was produced by Thurstone inasmuch as his goal was to illustrate a method of scaling, but other age maps might be constructed from the data. He acknowledged as secondary results a rapidly accelerating curve for intelligence in children/adolescents reaching a limit at about age twenty, with distributions of intelligence consistent for each age group from three to fourteen years of age.

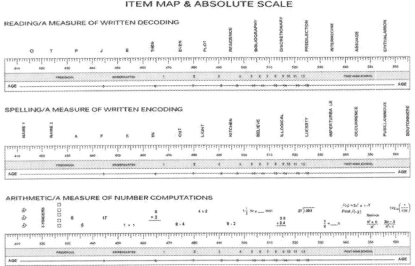

Figure 3.8 Wide Range Achievement Test: WRAT-3
From Wilkinson, G. S. 1993. *The Wide Range Achievement Test* (1993 Edition).
Wilmington, DE: Wide Range.

Figure 3.8 is a more contemporary map of an achievement variable: WRAT-3 (Wilkinson 1993; Stone 1995). This test of achievement measures (1) word naming, (2) arithmetic computation, and (3) spelling from dictation. Items are arranged according to difficulty. Their calibrations were used to map word reading, arithmetic computation, and spelling by dictation. These maps progress from left to right indicating increases in item difficulty and person ability. The arrangement of items suggests the expected arrangement of persons according to their abilities. Less able persons will be located to the left of persons more able. Able persons will find the easier items on the left less difficult than those items farther along to the right. These three variables follow developmental lines of learning corresponding to instructional goals, making test administration efficient and informative. The map of each WRAT-3 variable gives sample items illustrating progressive difficulty. Below the items is an equal-interval scale providing measures. The location of average grade level is given, as well as the average age associated with the item calibrations and person measures.

These psychometric maps have immediate application. Like the marks that are made on a doorjamb to show the increasing height of a child, this achievement map can show student progress on three scales. The maps picture order in items and in measures. Progress of pupils along the ruler can be indicated and supplemented by criterion and normative locations. The grade and age norms show expected growth along the variable, and the scale gives measures that are useful for interpretation.

Figure 3.9 The Lexile Framework for Reading
From Stenner, A. J. 1995. *Lexile Framework*. Research Triangle Park, NC: MetaMetrics.

Figure 3.9 is a reduced map of the Lexile Framework for Reading (copyright 1995, MetaMetrics). The master map is larger, more comprehensive, and requires a chart greater than 2' by 3' to picture only a small portion of the large amount of available information. Lexile calibration values have been computed for substantial numbers of trade books, texts, and tests. The title column indicates the content validity of the scaling. The educational levels column shows the increase in difficulty corresponding to reading more difficult materials.

Rasch mapping technology offers a powerful tool for conjointly ordering objects of measurement (i.e., readers) and indicants (i.e., texts). Meaning accrues as the conjoint ordering of reader and text is juxtaposed with other orderings, including grade level, income, or job classification. The figure below outlines the reading demands of specific careers.

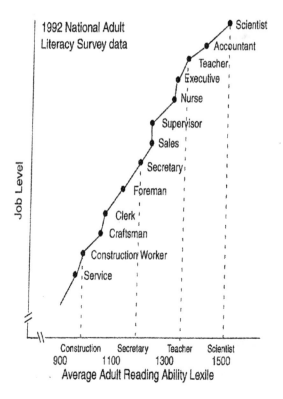

Collections of these person/item "orderings" constitute a rich interpretive framework for bringing meaning to the measurement of human behavior.

A significant leap in understanding and utility is accomplished when the ordering of indicants along the line of the variable can be predicted from theory. In every application of physical science measurement, instrument calibration is accomplished via theory, not data. Social science measurement stands alone in its reliance on data in the construction of instrument calibrations and cocalibrations between instruments.

Perhaps the key advantage of theory-based calibrations is that an absolute framework for measure interpretation can be constructed without reference to any individual or group measures on objects or indicants. The prospect of absolute measurement (i.e., not relative), long taken for granted in the physical sciences, has until recently eluded social scientists. The building of maps for the major dimensions of human behavior is now possible because of the theoretical work of Rasch and Wright, amplified by the work of colleagues.

One pretender to the kind of mapping process outlined above is evident in NAEP's use of the Reading Proficiency Scale (RPS). The RPS is a transformed Rasch scale with an operating range of 0 to 500. NAEP describes performance at grades 4, 8, and 12 as rudimentary, basic, intermediate, adept or advanced depending upon the RPS attained by each student. Thus, a rudimentary reader has an RPS = 150, an intermediate reader at RPS = 250, and an advanced reader at RPS = 350. These logical steps simply "name" certain "anchor" points on the RPS scale.

Problems develop when reader performance on the RPS scale is described using relative language such as a rudimentary reader can "follow brief written directions" or "can carry out simple, discrete reading tasks," or a basic reader can "understand specific or sequentially related information." An intermediate reader can "search for specific information, interrelate ideas, and make generalizations." An adept reader can "analyze and integrate less familiar material and provide reactions to and explanations of the text as a whole." An advanced reader can "understand the links between ideas even when those links are not explicitly stated."

These statements are not appropriate descriptions of scale points along the RPS scale; rather they are good description of the behavioral consequences of more or less accurately matching the demands of a text with the capabilities of a reader. Thus, rather than describing only absolute scale positions, these annotations describe differences between a reader measure and a text measure. When a text measure exceeds a reader's measure, comprehension is low, and the kinds of reader behaviors used above to describe rudimentary and basic results would be descriptive of that specific encounter between reader and text.

When a reader's measure exceeds a text measure, the kinds of reader behaviors used to describe adept and advanced readers are evident. The key point is that each of these behaviors can be elicited in the same reader simply by altering the level of text that is presented to the reader. Thus we can make a 400L (second grade level) reader adept by presenting a 100L text or a 400L reader rudimentary by presenting 800L text. Comprehension rate is always relative to the match between reader and text, and it is this rate rather than the reader's measure that is appropriately described in behavioral and proficiency terms. Much confusion has resulted from a failure to recognize this distinction.

Such maps meet and exceed the requirements of the Standards (1999) offered to guide test developers and consumers. Rather than being bogged down by verbiage, a well-documented psychometric map offers the essential information in a useful visual presentation. So-called content validity is easy to observe. So-called construct validity can be demonstrated by map locations for items, persons, and criteria exemplified by educational levels, ages, and other information positioned on the Lexile map. So-called criterion validity is directly observed by the additional data provided. We do not agree with the proliferation of flavors of validity. Validity is a simple idea. Does the measurement measure the intended construct? Types of validity are actually aspects of scale utility, not measurement validity.

Summary

Successful item calibration and person measurement can be used to produce a map of the variable. The resulting map is no less a ruler than the ones constructed to measure length. The realized map portrays the content, criterion, and construct utility of the variable. The empirical calibration of items and the measures of persons should correspond to the original intent of person and item placement, but changes must be made when correspondence is not achieved. There should be continuous dialogue between the plan, person measures, and item calibrations. Variables are never created once and for all. Continuous monitoring of the variable is required in order to keep the map coherent and up to date. Support for reliability and validity does not rest in once and for all coefficients but in substantiating relevant and stable indices for instruments and measures. Such indications must be continuously monitored in order to maintain the variable map and assure its relevancy. We conclude with a flowchart for the process:

Flowchart of Mapping:

1. Intention

    Initial Map of Variable

    Item Calibration with SE

2. Dialectic: Idea vs. Initial Map

    Item Fit Poor

        1. Eliminate Item(s)
        2. Revise Item(s)
        3. Add New Item(s)

    Item Fit Good

3.   Realization   Revised Map of Variable

    Make Measures with SE

1. The "intended map" of the variable is the idea and plan of our intentions.

2. There is continuous dialogue between the intention (idea) and realization (data) to maintain validity and quality control. The degree of correspondence between the maps of intention and realization indicates the degree of success achieved.

3. The "realized map" of the variable conveys in the item calibrations and person measures the best outcome to date while recognizing the process is ongoing.

Box 7: Descriptive Rasch Model versus Causation

Andrich (2004) makes the case that Rasch models are powerful tools precisely because they are prescriptive, not descriptive, and when model prescriptions meet data, anomalies arise. Rasch models invert the traditional statistical data-model relationship. Rasch models state a set of requirements that data must meet if those data are to be useful in making measurements. These model requirements are independent of the data. It does not matter if the data are bar presses, counts correct on a reading test, or wine taste preferences. If these data are to be useful in making measures of rat perseverance, reading ability, or vintage quality, all three sets of data must conform to the same invariance requirements. When data fail to fit a model, Rasch measurement theory (Rasch 1960; Andrich 1988, 2010; Wright 1977, 1999) does not respond by relaxing the invariance requirements and adding, say, an item specific discrimination parameter to improve fit, as does Item Response Theory (Hambleton, Swaminathan, and Rogers 1991). Rather, the Rasch approach is to examine the items serving as the medium for making observations, and to change them in ways likely to produce new data conforming with the modeled expectations.

A causal Rasch model (in which item calibrations come from theory, not data) is then doubly prescriptive (Stenner, Stone, and Burdick 2009). First, in accord with Rasch, it is prescriptive regarding the data structures that must be present:

> The comparison between two stimuli should be independent of which particular individuals were instrumental for the comparison; and it should also be independent of which other stimuli within the considered class were or might also have been compared. Symmetrically, a comparison between two individuals should be independent of which particular stimuli within the class considered were instrumental for comparison; and it should also be independent of which other individuals were also compared, on the same or on some other occasion. (Rasch 1961)

Second, causal Rasch models (Burdick, Stone, and Stenner 2006; Stenner, Burdick, and Stone 2008) prescribe the values imposed by substantive theory on the item calibration estimates. Thus, the data, to be useful in making measures, must conform to both Rasch model invariance requirements *and* to substantive theory invariance requirements as specified by the theoretical item calibrations.

When data meet both sets of requirements, then those data are useful not just for making measures of some vaguely defined construct but are useful for making measures of that precise construct specified by the equation that produced the theoretical item calibrations. We note again that these dual invariance requirements come into stark relief in the extreme case of no connectivity across stimuli or examinees ($x \; i \; p$). How, for example, are two readers to be measured on the same scale if they share no common text passages or items? If you read *The Hunger Games* and answer machine-generated questions about it, and I read *Lord of the Rings* and answer machine-generated questions about it, how would it be possible to realize an invariant comparison of our reading abilities except by means of predictive theory? How else would it be possible to know that you read 250L better than I, and, furthermore, that you comprehended 95 percent of what you read, whereas I comprehended 75 percent of what I read? Most importantly, by what other means than theory would it ever be possible to reproduce this result to within a small range of error using another two completely different books as the basis of comparison?

Given that seemingly nothing is in common between the above two reading experiences, invariant comparisons might be thought impossible. Yet in the thermometer example, it is in fact a routine everyday experience for different instruments to be interpreted as informing comparable measures of temperature. Why are we so quick to accept that you have a high-grade fever of 104°F, and I have a low-grade fever of 100°F (based on measurements from two different thermometers), and yet find the book reading example inexplicable, even unacceptable to some people? The answer lies in well-developed construct theory, rigorously established instrument engineering principles, and uniform metrological conventions (W. Fisher 2009).

Clearly, each of us has had ample confirmation that weight denominated in pounds and kilograms can be well measured by any reputable manufacturer's bathroom scale. Experience with diverse bathroom scales has convinced us that, within a pound or two of error, these instruments will produce not just invariant relative differences between two persons but will also meet the more stringent expectation of invariant absolute magnitudes for each individual independent of instrument. Over centuries, instrument engineering has steadily improved to the point that for most purposes, "uncertainty of measurement" (usually interpreted as the standard deviation of a distribution of imagined or actual replications taken on a single person) can be effectively ignored for most bathroom scale applications. And, quite importantly, by convention (i.e., the written or unwritten practice of a community), weight is denominated in standardized units (kilograms or pounds). The choice of any given unit is arbitrary, but what is decisive is that a unit is agreed to by the community and is maintained through consistent implementation, instrument manufacture, and reporting. At present, language ability (reading, writing, speaking, listening) does not enjoy a common construct definition, nor a widely promulgated set of instrument specifications, nor a conventionally accepted unit of measurement. The challenges that must be addressed in defining constructs, specifying instrument characteristics, and standardizing units include cultural assumptions about number and objectivity, political challenges in shaping legislation, resource allocation, and the expectations and procedures of social scientists (W. Fisher 2009, 2012a). In this context, the Lexile Framework for Reading (Stenner, H. Burdick, Sanford, and D. Burdick 2006) stands as an exemplar of how psychosocial measurement can be unified in a manner precisely parallel to the way unification was achieved for length, temperature, weight, and dozens of other useful attributes (Stenner and Stone 2010).

A causal (constrained) Rasch model (Stenner, Stone, and Burdick 2009) that fuses a substantive theory to a set of axioms for conjoint additive measurement affords a much richer context for the identification and interpretation of anomalies than does a descriptive (i.e., unconstrained) Rasch model. First, with the measurement model and the substantive theory fixed, anomalies are understood as problems with the data. Attending to the data ideally leads to improved observation models (e.g., new task types) that reduce unintended dependencies and variability. An example of this kind of improvement in measurement was realized when the duke of Tuscany put a top on some of the early thermometers, thus reducing the contaminating influences of barometric pressure on the measurement of temperature. In contrast with the descriptive paradigm dominating much of education science, the duke did not propose parameterizing barometric pressure in the model in the hope that the boiling point of water at sea level, as measured by open top thermoscopes, would then match the model expectations at three thousand feet above sea level.

Second, with both model and construct theory fixed, our task is to produce measurement outcomes that fit the invariance requirements of both measurement theory and construct theory. By analogy, not all fluids are ideal as thermometric fluids. Water, for example, is nonmonotonic in its expansion with increasing temperature and is especially unpredictable near its freezing point. Mercury, by contrast, has many useful properties as a thermometric fluid. Does the discovery that not all fluids are useful thermometric fluids invalidate the concept of temperature? No! In fact, a single fluid with the necessary properties would suffice to validate temperature as a useful construct—hence, the ubiquity of the mercury thermometer. The existence of a persistent invariant framework makes it possible to identify anomalous behavior (water's strange behavior) and interpret it in an expanded theoretical framework (Chang [2004] 2007).

Analogously, finding that not all reading item types conform to the dual invariance requirements of a Rasch model and the Lexile theory does not invalidate either the axioms of conjoint measurement theory or the

Lexile reading theory. Rather, the anomalous behaviors of some kinds of text (recipes, calendars, poems) are open invitations to expand the theory to account for these deviations from expectation. Notice here the subtle shift in perspective. We do not need to find one thousand unicorns; one will suffice to establish the reality of the class. The finding that reader behavior on a minimum of two types of reading tasks can be regularized by the joint actions of the Lexile theory and a Rasch model is sufficient evidence for the reality of the reading construct. Of course, actualizing this scientific reality to make the reading construct a universally uniform and available object in the world will require the investment of significant social, legal, and economic resources (W. Fisher 2009; W. Fisher and Stenner 2012).

Equation (1) is a causal Rasch model for dichotomous data, which sets a measurement outcome (expected score) equal to a sum of modeled probabilities:

$$Expected\ raw\ score =: \sum_{i} \frac{e^{(b-di)}}{1 + e^{(b-di)}}$$

$$(1)$$

The measurement outcome is the dependent variable, and the measure (e.g., person parameter, b) and instrument (e.g., the parameters d_i pertaining to the difficulty d of item i) are independent variables. The measurement outcome (e.g., count correct on a reading test) is observed, whereas the measure and instrument parameters are not observed but can be estimated from the response data and substantive theory, respectively. When an interpretation invoking a predictive mechanism is imposed on the equation, the right-side variables are presumed to characterize the process that generates the measurement outcome on the left side. The symbol =: was proposed by Euler circa 1734 to distinguish an algebraic identity from a causal identity (right-hand side causes the left-hand side). This symbol (=:) was reintroduced by Judea Pearl and can be read as indicating that manipulation of the right-hand side via experimental intervention will cause the prescribed change in the left-hand side of the equation. But simple use of an equality (=) does not signal a causal interpretation of the equation.

A Rasch model combined with a substantive theory embodied in a specification equation provides a more or less complete explanation of how a measurement instrument works (Stenner, Stone, and Burdick 2009). A Rasch model in the absence of a specified measurement mechanism is merely a probability model. A probability model absent a theory may be useful for describing or summarizing a body of data, and for predicting the left side of the equation from the right side, but a Rasch model in which instrument calibrations come from a substantive theory that specifies how the instrument works is a causal model. That is, it enables prediction after intervention.

Below we summarize two key distinguishing features of causal Rasch models and highlight how these features can contribute to improved ENL, EFL, and ESL measurement.

1. Causal Rasch models are individually centered, meaning that a person's measure is estimated without recourse to data on other individuals. The measurement mechanism that transmits variation in the language attribute (within person over time) to the measurement outcome (count correct on a reading test) is hypothesized to function the same way for every person. This hypothesis is testable at the individual level using Rasch Model fit statistics.

2. Figuring prominently in the measurement mechanism for language measurement is text complexity. The specification equation used to measure text complexity is hypothesized to function the same way for most text genres and for readers who are ENL, EFL, and ESL. This hypothesis is also testable at the individual level, but aggregations can be made to examine invariance over text types and reader characteristics.

Point one emphasizes that individual measurement is desired and should be the major focus of all assessment. Without this emphasis on the individual, there can be no assessment of growth or progress, no assemblage of group findings. Point two emphasizes the need for a substantive consideration in all measurement inquiries, without which assessment cannot proceed.

The data for computing empirical text complexity measures came from the reader appliance in EdSphere. Students access tens of millions of professionally authored digital texts by opening EdSphere and clicking on the Reader App. Digital articles are drawn from hundreds of periodicals, including *Highlights for Children, Boys' Life, Girls' Life, Sports Illustrated, Newsweek, Discovery, Science, The Economist, Scientific American,* and so forth. Such a large repository of high-quality informational text is required to immerse students with widely varying reading abilities in daily deliberate practice across the K–16 education experience.

Students use three search strategies to locate articles targeted at their Lexile level: (1) click on suggested topics, (2) click the icon "Surprise Me," or the most frequently used method, (3) type search terms into "Find a Book or Article" (see figure B7.1). In the example below, a 1069L reader typed "climate change" in the search box and found 13,304 articles close to her reading level. The first article is an 1100L four-page article from *Scientific American* with a short abstract. Readers browse the abstracts and refine the search terms until they find an appropriate length article about their interest topic (or a teacher-assigned topic) at their reading level.

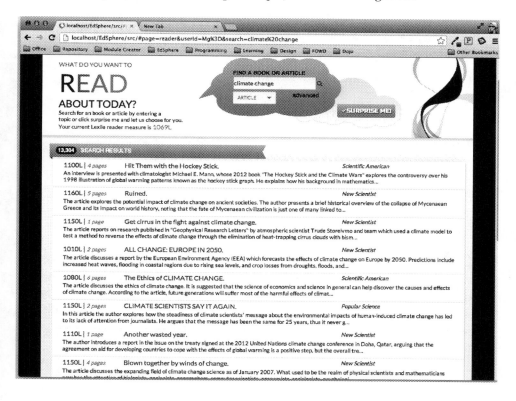

Figure B7.1. Results from student's keyword search for articles about climate change; results include publication titles, publication dates, and (or) page length

Within one second of selecting an article, the machine builds a set of embedded auto-semantic cloze items. In the example below, a student selected the article "A Low-Carbon Future Starts Here" (see figure B7.2). The auto-generated semantic cloze engine created nineteen cloze items for this article (see upper right-hand corner). Students choose from the four options that appear at the bottom of the page. The incorrect options have similar difficulty and part of speech to the correct answer. The answer is autoscored, and the correct answer is immediately restored in the text and color coded as to whether the student answered correctly or incorrectly. The goal is to read-to-learn so the system will not let an incorrect choice undermine comprehension.

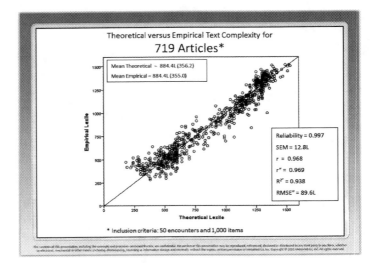

Three instructional supports are built into the Reader App to facilitate comprehension. *First,* suggested strategies are presented to students during the reading process. *Second,* students have access to an in-line dictionary and thesaurus (one-click access). Finally, a text-to-speech engine has been integrated into EdSphere, allowing words, phrases, or sentences to be machine-spoken to the reader.

Figure B7.3 presents the results of a multiyear study of the relationship between theoretical text complexity as measured by the Lexile Analyzer (freely available for noncommercial use at Lexile.com) and empirical text complexity as measured by the EdSphere platform. Each of the 719 articles included in this study was evaluated by the analyzer for semantic demand (log transformed frequency of each word's appearance in a multibillion-word corpus) and syntactic demand (log transformed mean sentence length). The text preprocessing, what constitutes a *word*, involves thousands of lines of code. Modern computing enables the measurement of the Bible in a couple of seconds.

The EdSphere platform enables students to select articles of their choosing from a collection of over one hundred million articles that have been published and measured over the past twenty years. As a student's reading ability grows, a 200L window moves up the scale (100L below the student's ability to 100L above), and all articles relevant to a reader's search term that have text complexity measures in the window are returned to the reader. The machine generates a four-choice cloze every seventy to eighty words, and the count correct combined with the reader's Lexile measure is used to compute an empirical text complexity for the article averaged over at least fifty readers and at least a thousand items.

The 719 articles chosen for this study were the first articles to meet the dual requirements of at least fifty readers and at least a thousand item responses. Well-estimated reader measures were available prior to the encounter between an article and a reader. Thus, each of the articles has a theoretical text complexity measure from the Lexile Analyzer and an empirical text complexity from EdSphere. The correlation between theory and empirical text complexity is r = 0.968 (r² = .938).

Connecting the Psychometrics to an Instructional Theory for Language Development

Work on connecting causal Rasch models to theories of language development (Hanlon 2013; Swartz, Hanlon, Stenner, and Childress 2015) has made extensive use of Ericsson's theory of deliberate practice in the acquisition of language expertise. Deliberate practice is a core tenant of Ericsson's theory of expertise development. Hanlon (2013) distills five core principles of deliberate practice in the development of reading ability: (1) targeted practice reading text that is not too easy and not too hard, (2) real-time corrective feedback on embedded response requirements, (3) distributed practice over a long period of time (years, decades), (4) intensive practice that avoids burnout, and (5) self-directed options when one-on-one coaching is not available. Each of these principles, when embedded into instructional technologies, benefits from individually centered psychometric models in which, for example, readers and text are measured in a common unit.

Swartz, Hanlon, Stenner, and Childress (2015) provide a complete description of EdSphere, its history, and components. The EdSphere Technology is designed to immerse students in deliberate practice in reading, writing, content vocabulary, and practice with conventions of standard English:

> These principles of deliberate practice are strengthened by embedding psychometrically sound assessment approaches into learning activities. For example, students

respond to cloze items while reading, compose short and long constructed responses in response to prompts, correct different kinds of convention errors (i.e., spelling, grammar, punctuation, capitalization) in authentic text, and select words with common meanings from a Thesaurus-based activity. Each item encountered by students is auto-generated and auto-scored by software. The results of these learning embedded assessments are especially beneficial when assessment item types are linked to a developmental scale. (Swartz, Hanlon, Stenner, and Childress 2015)

Figure B7.3 is an individual-centered reading growth trajectory denominated in Lexiles. All data comes from EdSphere. Student 1528 is an ESL seventh-grade male (first language Spanish) who read 347 articles of his choosing (138,695 words) between May 2007 and April 2011. Each solid dot corresponds to a monthly average Lexile measure. The growth curve fits the monthly means quite well, and this young man is forecasted (big dot on the far right of the figure) to be a college-ready reader when he graduates from high school in 2016. The open dots distributed around zero on the horizontal axis are the expected performance minus observed performance (in percentages) for each month. Expected performance is computed using the Rasch model and inputs for each article's text complexity and the updated reader's ability measure. Given these inputs, EdSphere forecasts a percent correct for each article encounter. The observed performance is the observed percentage correct for the month. The difference between what the substantive theory (Lexile Reading Framework) in cooperation with the Rasch model expects and what is actually observed is plotted by month. The upper left-hand corner of the graphic summarizes the expected percentage correct over the four years (73.5 percent) and observed percentage correct (71.7 percent) across the 3,342 items taken by this reader. Note that EdSphere is dynamically matching text complexity of the articles the reader can choose to the increasing reader ability over time. So, this graphic describes a within-person (intraindividual) test of the quantitative hypothesis: can EdSphere trade off a change in reader ability for a change in text complexity to hold constant the success rate (comprehension)? For this reader, the answer appears to be a resounding yes!

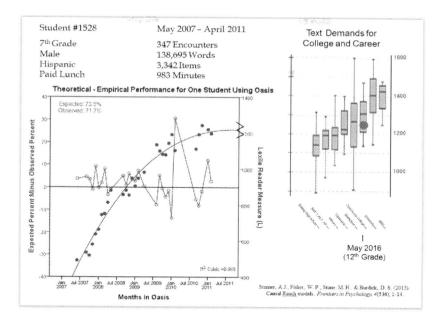

Figure B7.4 is a graphical depiction of the 99 percent confidence interval for the artifact corrected correlation between theoretical and empirical text complexity. The artifacts included measurement error, double range restriction, and construct invalidity. The artifact-corrected correlation (coefficient of theoretical equivalence) is slightly higher than r = 1.0, suggesting that the Lexile theory accounts for all true score variation in the empirical text complexity measures. The reader may be puzzled about how a correlation can be higher than r = 1.0; of course it can't be, but an artifact-corrected correlation can be if the artifactors used in the correction are, perhaps due to sampling error, lower than their population values.

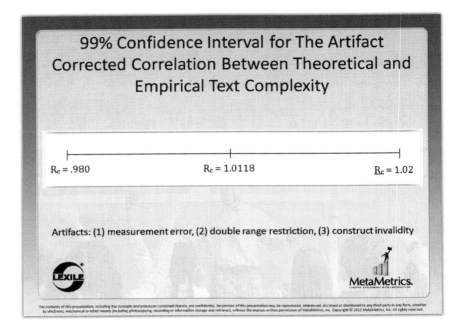

Notes

1.   The NexTemp thermometer is a small plastic strip pocked with multiple enclosed cavities. In the Fahrenheit version, forty-five cavities arranged in a double matrix serve as the functioning end of the unit. Spaced at 0.2°F intervals, the cavities cover a range from 96.0°F to 104.8°F. Each cavity contains three cholesteric liquid crystal compounds and a soluble additive. Together, this chemical composition provides discrete and repeatable change-of-state temperatures consistent with the device's numeric indicators. Change of state is displayed optically (cavities turn from green to black) and is easily read.

2.   Text complexity is predicted from a construct specification equation incorporating sentence length and word frequency components. The squared correlation of observed and predicted item calibrations across hundreds of tests and millions of students over the last fifteen years averages about .93. Recently available technology for measuring reading ability employs computer-generated items built on the fly for any continuous prose text in a manner similar to that described for mathematics items

by Bejar and colleagues (Bejar et al. Revuelta 2003). Counts correct are converted into Lexile measures via a Rasch model estimation algorithm employing theory-based calibrations. The Lexile text measure of the target text and the expected spread of the cloze items are given by theory and associated equations. Differences between two readers' measures can be traded off for a difference in Lexile text measures. When the item generation protocol is uniformly applied, the only active ingredient in the measurement mechanism is the choice of text complexity (choosing a 500L or 1000L article, for example).

In the temperature example, a uniform increase or decrease in the amount of soluble additive in each cavity changes the correspondence table that links the number of cavities that turn black to degrees Fahrenheit. Similarly, an increase or decrease in the text demand (Lexile) of the passages used to build reading tests predictably alters the correspondence table that links count correct to Lexile reader measure. In the former case, a temperature theory that works in cooperation with a Guttman model produces temperature measures. In the latter case, a reading theory that works in cooperation with a Rasch model produces reader measures. In both cases, the measurement mechanism is well understood, and we exploit this understanding to address a vast array of counterfactuals (Woodward 2003). If things had been different (with the instrument or object of measurement), we could still answer the question as to what then would have happened to what we observe (i.e., the measurement outcome). It is this kind of relation that illustrates the meaning of the expression "There is nothing so practical as a good theory" (Lewin 1951).

Box 8: Developmental Science

Hasok Chang in his book *Inventing Temperature* ([2004] 2007) coins the phrase *epistemic iteration* (234) to signify the continuous progress of scientific investigation and especially the self-correction process inherent in good science. This does not mean that mistakes are not made but that all scientific announcements need to be verified, something not evident in today's pop media. This concept is not new. Newton's commonly cited phrase "If I have seen further, it is by standing on the shoulders of giants" expresses the process metaphorically. An earlier expression attributed to Bernard of Chartres by John of Salisbury states, "We are like dwarves perched on the shoulders of giants, and thus we are able to see more and farther than they were able."

Thomas Kuhn's landmark book *The Structure of Scientific Revolutions*

(1962) is well known and referenced frequently. Less known is a small paperback by Nobel laureate James Conant, *On Understanding Science* (1947), published fifteen years earlier. Kuhn acknowledges the mentorship of Conant in his preface to TSSR and also in the preface of an earlier work, *The Copernican Revolution* (1956). Conant's book stems from an earlier popular science lecture he delivered at Yale, expressing a clear message for promoting what he designates "The Tactics and Strategies of Science." Albeit in a simplified expression, Conant preempts Kuhn's message by more than fifteen years! He advocates science teachers giving less attention to the "fruits of science" (i.e., the outcome of facts) and more attention to "the ways in which these fruits have been attained" (33). He deduces, "Science emerges from the other progressive activities of man to the extent that new concepts arise from experiments and observations and the new concepts in turn lead to further experiments and observations" (37).

Conant discusses two critical developments to illustrate his point in "The Tactics and Strategies of Science." The first matter concerns "air" stemming from investigations by Galileo to Boyle. The second matter is "phlogiston," from which we draw the essential points made by Conant.

> It is sometimes said that the experimenters before Lavoisier's day did not carry out quantitative experiments, that is, they did not use the balance. If they had we are told, they would have discovered that combustion involves an increase in weight and would have rejected the phlogiston theory. This is nonsense. Rey, as I have already explained, long before the beginning of the phlogiston period showed that a calx [oxide today] weighed more than a metal. Quantitative experiments, though, of course, not very accurate ones, were repeatedly made. Everyone knew that a calx weighed more than the metal from which it was formed. No straightforward statement of the phlogiston theory could accommodate this fact. Yet the phlogiston theory was so useful that few if any in the mid-eighteenth century were looking to overthrow it or disapprove it. Rather, they were interested in reconciling one inconvenient set of facts with what seemed from their point of view an otherwise admirable conceptual scheme. How they twisted and squirmed to accommodate the quantitative facts of calcination with the phlogiston theory makes an interesting chapter in the history of science. (94)

> The principle which emerges is one already encountered, namely, that it takes a new conceptual scheme to cause the abandonment of an old one; when only a few facts appear to be irreconcilable with a well-established conceptual scheme, the first attempt is not to discard the scheme, but to find some way out of the difficulty and keep it. (94)

Our strategy has been to use expressions from engineering and manufacturing to signify the processes of development and evolution in measuring (i.e., production following upon a substantive-based theory), hence, *theory-driven constructed measures*.

Herbert Simon (1969) distinguished between two formats or approaches to conducting science, which he designated as (1) *natural science* or "knowledge about natural objects and phenomena," and (2) *artificial science* (i.e., fabricated), "meaning man-made as opposed to natural" (4). Throughout his book, Simon uses terminology such as engineering, fabricating, producing, and manufacturing. "It has been the task of the science disciplines to teach about natural things: how they are and how they work. It has been the task of engineering school to teach about artificial things: how to make artifacts that have desired properties and how to design" (56).

Artifact is a key word for our approach to what Simon means by "artificial science." The history of length, time, and temperature illustrate the fabrication of what we designate *measuring mechanisms* (Stenner, Stone, and Burdick 2009). Building a measuring mechanism constructs an artifact that produces measures. The process is developmental, requiring continual refinement and continuous quality control, which is our approach to a constant iterative process.

Box 9: A Significant Difference from 1.0

There have been numerous critiques of significance testing, and we do not wish to add to them except to make our point. Why test for a difference from zero when the goal of significance testing is to demonstrate the hypothesis of a substantive theory? Those researchers who celebrate when they reach some modest level of significance are at the wrong end of the index. If a hypothesis is to exhibit any practical value, it should demonstrate its importance by achieving a level of 1.0 (within the confidence level). This "significance" test (if it is to be conducted) should evaluate how far the researcher's hypothesis is from where it was posited to be; otherwise the hypothesis is suspect even before the investigation begins.

A significance test evaluates the relationship between the dependent variable and independent variable(s). A result validating the hypothesis should exhibit a correlation at the high end of the scale, not simply a significant difference from zero (the low end of the scale). This is less an indictment of significance testing than it is a criticism of the quality of the hypothesis and that of the originator.

An example makes the point.

Box 10: Perspectives on Reliability and Validity

Working from the assumption that these matters are important in measuring, our question is how they are to be defined, applied, and interpreted. Numerous critiques have been made, positions taken, and comprehensive reviews written. There is no need to attend to all that has been covered elsewhere by books, chapters, and articles; our goal here is simply stating our position as to their influence on our work.

Reliability seems a straightforward principle: a measuring instrument should provide almost exact values, the same way a recipe should produce a predictable outcome in cooking. Manufacturing requires that products meet the design standards and tolerance specifications in the construction, and continuous quality control monitors these specifications. Fashioning a test requires attention to these same criteria. An outcome is either by chance or design. The design provides the plain language goal, the formal statement, and the specification equation of the instrument. These are the essentials for judging reliability. Boxes 3, 6, 8, and 9 address these matters as well.

Perspectives on validity tend to waver between the (now classic) four types or else call for their unity. Chapter 11, "Validity Is Theoretical Equivalence," provides the basis for our views. Validity should supply confirmation of our theory: we construct a theory of measures and then must demonstrate its occurrence. Boxes 7, 8, 9, and 10 speak to these points.

Louis Guttman (1976, 26, 82–107), in a paper in the *Statistician* entitled "What Is Not What in Statistics," clearly delineated the problems with the blind interpretation of significance tests and misuse of *p*-value? It is helpful to state his position in a sequence of quotes:

> *The symbols of inference rather than the substance may have taken over.* This appears to be especially true in the social

sciences ... Referees and editors of some journals insist on decorating tables of various kinds of data with stars and double stars, and on presenting lists of 'standard errors' despite the fact that the implied probabilities for significance or confidence are quite erroneous from the point of view of statistical inference. (82)

Many if not most practitioners do not do the scientific thinking that must precede statistical inference. They do not make the choice of null versus alternative hypothesis that is properly tailor-made to their specific substantive problem. *They behave as if under the delusion that the choice is not in their hands, that the null hypothesis is pre-determined either by the mathematicians who created modern statistical inference or by some immutable and content less principles of parsimony.* (82)

Most social science inference problems are multivariate at the outset, yet they are usually not studied as such. (85)

Both estimation and the testing of hypotheses have usually been restricted as if to one-time experimentation, both in theory and in practice. But the essence of science is replication: a scientist should always be concerned about what will happen when he or another scientist repeats his experiment. (86)

Suppose a scientist rejects null hypotheses in favor of a given alternative: what is the probability that the next scientist's experiment will do the same? Merely knowing the probabilities for type I and type II errors of the first experiment is not sufficient for answering the question ... *Logically, should not the original alternative hypothesis become the null hypothesis for the second experiment?* (86)

Without substantive knowledge of their respective fields, there is no basis for assigning roles to hypotheses as 'null' or 'alternative'? A first approximation is not the null hypothesis talked about in textbooks. (88)

Orthogonality. The question for 'independent' contributions from each of several correlated components is a perennial enterprise of non-mathematicians ...

Mathematicians know that each such orthogonality is but an artifact created by the designer of experiments, and may have nothing to do with the inter-relation of natural phenomenon ... Most contexts in which orthogonality occurs in statistics are created by the statistical analysis, and that *orthogonality has no necessary 'natural' interpretation or implication ...* Generally predictors correlate with each other, and there is no intrinsic statistical bootstrap operation for designing 'independent' contribution in this case. Even for the case of uncorrelated predictors, there is no guarantee that a further predictor cannot be found that will correlate with the old, restoring the impossibility of giving independent credit to each of the predictors separately. (88–89)

The associated techniques for data analysis cannot presume to be amenable to 'exact' tests of significance, whether non-parametric or parametric. Indeed, they suggest looking at inference itself; why should one be interested in an 'exact' level of significance or confidence? *Non-inferential data analysis is content with being descriptive, and often only with a 'first approximation' with some indication of how approximately it is exact ... Replication is the test of science, and repeated replications – however approximate – may be worth more than trying to assess the 'exactness' from a level of approximation of one or two trials.* (89)

Our emphasis is given in italics. We subscribe to these interpretations by Guttman. Replication is absolutely necessary, and it must be continuously addressed.

Box 11: Individual-Centered versus Group-Centered Measures

The preface to Probabilistic Models for Some Intelligence and Attainment Tests (Rasch 1960) cites Skinner (1956) and Zubin (1955). In an argument whereby, "... individual-centered statistical techniques require models in which each individual is characterized separately and from which, given adequate data, the individual parameters can be estimated" (Rasch 1960, xx). The Skinner reference is easily located. The mimeographed work by Zubin has not been found, but we did find another Zubin paper given at the 1955 ETS Invitational Conference on testing problems in which he writes, "An Example of the application of individual-centered techniques which keeps the sights of the experimenter focused on the individual instead of on the group ..." (116) may have helped Rasch situate his thinking. Rasch goes on to state, "... present day statistical methods are entirely group-centered so that there is a real need for developing individual-centered statistics" (xx). What constitutes the differences in these statistics?

While it is individual persons and groups of persons that are the focus of discussion, we begin with an even more simple illustration because human behavior is complex, and a single mechanical-like variable is a better illustration to one that is complex. We choose temperature for this illustration because measuring mechanisms (Stenner, Stone, and Burdick 2004) for temperature are well established and all report out in a common metric or degree (disregarding windchill, etc.). A measuring mechanism consists of (1) guiding substantive theory, (2) successful instrument fabrication, and (3) demonstrable data by which the instrument has established utility in the course of its developmental history.

Consider six mercury-tube outdoor thermometers that are placed appropriately in a local environment but near each other. They all register approximately the same degree of temperature, independently verified by consulting NOAA for the temperature at this location. One by one, each thermometer is placed in a compartment able to increase/decrease the prevailing temperature by at least ten degrees. Upon verifying the artificially induced temperature change for each thermometer, it is returned to its original location and checked to see if it returns to its previous value and agrees with the other five.

If each of the six thermometers measured a similar and consistent degree of temperature before and after the induced environmental intervention/manipulation, this consistency of instrument recording validates a deep understanding of the attribute "temperature" and its measurement. Each thermometer initially recorded the same temperature and, following a change to and from the artificial environment, returned to the base degree of temperature. Furthermore, all the measurements agree.

Interestingly, the experimentally induced change of environment also produced what may be called *causal validity*, not unlike construct validity (Cronbach and Meehl 1954) inasmuch as the temperature was manipulated, fabricated, engineered, and so forth via construction and use of the artificial environment. When measuring mechanism(s) such as outdoor thermometers are properly manufactured, this result is to be expected, and this experimental outcome and its replication would be predicted prior to environmental manipulation from all we know about temperature and thermometers. This outcome might further be termed *validity as theoretical equivalence* (Lumsden and Ross 1973) because the replications produced by all six thermometer recordings might be considered "one" temperature. Our theoretical prediction is expected as a consequence of the causal process produced by the experiment and reported by all the instruments. *Causal validity* is a consequence of the successful theoretical predictions realized in the experiment. Its essence is "prediction under intervention." The manipulable characteristics of our experiment involving the base environment, change made by way of an artificial environment, and the final change of recorded temperature are the consequence of a well-functioning construct theory and measuring mechanism. Each of the six individual thermometers records a similarly induced experimental deviation and a return to the base state. Each thermometer constitutes an individual unit, and the six thermometers constitute a group, albeit without variation, which is exactly what would be predicted.

Now consider a transition to human behavior. Space is the new outcome measure, and the determination of height at a point in time can be obtained from another well-established measuring mechanism— the ruler, which provides a point estimate for one individual measured at a single point in time. When this process is continued for the same individual over successive time periods, we produce a trajectory of height for the person over time (purely individual centered, as no reference to any other person(s) is required). From these values, one may determine growth over time intervals as well as any observed plateaus and spurts well known to occur in individual development. The individual's trajectory rate may also vary because of illness and old age, so we could discover different rates over certain time periods as well as determine a curvilinear average to describe the person's life course trajectory. Growth in height is a function of time, and the human characteristics entailed in a person's overall development result from genetic and environmental makeup. These statistics are intra-individually determined. Such statistical analyses produce the "individual-centered statistics" that Rasch spoke about.

Aggregating individual measurements of height into a group or groups is a common method for producing "group-centered statistics" often employing some frequency model such as the normal curve. This is most common when generalizing the characteristics of human growth in overall height based upon a large number of individuals. The difference between measuring a group of individuals compared to our first illustration using a group of thermometers is that while we expected no deviation among the thermometers, we do not expect all individuals to gain the same height over time but rather to register individual differences. Hence, we resort to descriptive statistics to understand the central trend and the amount of variation found in the group or groups. An obvious group-centered statistical analysis might aggregate by gender, comparing the typical height of females to males, for example.

The measurement of height is straightforward, and the measurement mechanism has been established over several thousand years. The same cannot be said for measuring mental attributes occurring in psychological, health, and educational investigations. Determining the relevant characteristics for their measurement is more difficult, although the procedures for their determination should follow those already discussed. The major statistical hurdle is moving from the ordering of a variable's units to its "measurement application." The measurement models of Georg Rasch have been instrumental in driving this process forward.

Do we know enough about the measurement of reading that we can manipulate the comprehension rate experienced by a reader in a way that mimics the above temperature example? In the Lexile Framework for Reading (LFR), the difference between text complexity of an article and the reading ability of a person is causal on the success rate (i.e., count correct). It is true that short-term manipulation of a person's reading ability is, at present, not possible, but manipulation of text complexity is possible because we can select a new article that possesses the desired text complexity such that any difference value can be realized. Concretely, when a 700L reader encounters a 700L article, the forecasted comprehension rate is 75 percent. Selecting an article at 900L results in a decrease in forecasted comprehension rate to 50 percent. Selecting an article at 500L results in a forecasted comprehension rate of 90 percent. Thus we can increase/decrease comprehension rate by judicious manipulation of texts (i.e., we can experimentally induce a change in comprehension rate for any reader and then return the reader to the "base" rate of 75 percent). Furthermore, successful theoretical predictions following such interventions are invariant over a wide range of environmental conditions, including the demographics of the reader (male, adolescent, etc.) and the characteristics of text (length, topic/genre, etc.).

Many applications of Rasch models to human science data are thin on substantive theory. Rarely proposed is an a priori specification of the item calibrations (i.e., constrained models). Causal Rasch models (Stenner, Fisher, Stone, and Burdick 2013; Burdick, Stone, and Stenner 2006; Stenner, Stone, and Burdick 2009; Stenner and Stone 2010) prescribe (via engineering and manufacturing quality control) that item calibrations take the values imposed by a substantive theory. For data to be useful in making measures, those data must conform to the invariance requirements of both the Rasch model and the substantive theory. Thus, causal Rasch models are doubly prescriptive. When data meet both sets of requirements, the data are useful not just for making measures of some construct but are useful for making measures of that precise construct specified by the equation that produced the theoretical item calibrations.

A causal (doubly constrained) Rasch model that fuses a substantive theory to a set of axioms for conjoint additive measurement affords a much richer context for the identification and interpretation of anomalies than does an unconstrained descriptive Rasch model. First, with the measurement model and the substantive theory fixed, it is self-evident that anomalies are to be understood as problems with the data ideally leading to improved observation models that reduce unintended dependencies in the data (Andrich 2002). Second, with both model and construct theory fixed, it is obvious that our task is to produce measurement outcomes that fit the (aforementioned) dual invariance requirements. An unconstrained model cannot distinguish whether it is the model, data, or both that are suspect.

Over centuries, instrument engineering has steadily improved to the point that for most purposes "uncertainty of measurement," usually reported as the standard deviation of a distribution of imagined or actual replications taken on a single person, can be effectively ignored. The practical outcome of such successful engineering is that the "problem" of measurement error is virtually nonexistent; consider most bathroom scale applications. The use of pounds and ounces also becomes arbitrary, as is evident from the fact that most of the world has gone metric although other standards remain. What is decisive is that a unit is agreed to by the community and is slavishly maintained through substantive theory together with consistent implementation, instrument manufacture, and reporting. We specify these stages:

Theory → Engineering → Manufacturing → Quality Control

The doubly prescriptive Rasch model embodies this process. Different instruments *qua* experiences underlie every measuring mechanism: environmental temperature, human temperature, children's reported weight on a bathroom scale, reading ability. From these illustrations and many more like them, we determine point estimates and individual trajectories and group aggregations. This outcome lies in well-developed construct theory, instrument engineering, and manufacturing conventions that we designate *measuring mechanisms*.

4

DOES THE READER COMPREHEND THE TEXT BECAUSE THE READER IS ABLE OR BECAUSE THE TEXT IS EASY?

Rasch's unidimensional models for measurement are conjoint models that make it possible to put both texts and readers on the same scale. Causal Rasch models (Stenner, Fisher, Stone, and Burdick 2013) for language testing fuse a theory of text complexity to a Rasch model, making it possible for a computer to response illustrate texts read by language learners during daily practice. Causal Rasch models are doubly prescriptive. First, they are prescriptive as to data structure (e.g., noninteracting item characteristic curves). Second, they are prescriptive as to the requirements of a substantive theory. One consequence of this fusion of a Rasch model with a substantive theory is that individual-centered growth trajectories can be estimated for each reader even though no two readers ever read the same article or respond to a single common item. Rather than common items or common persons being the connective tissue that makes comparisons of readers possible, common theory is the connective tissue, just as is true in human temperature measurement where each person is paired with a unique thermometer. Thus, although the instrument is unique for each person on each occasion, a text complexity theory makes it possible to convert counts correct to a common reading ability metric in each and every application.

A new paradigm for measurement in education and psychology, which mimics much more closely what goes on in physical science, was foreshadowed by Thurstone (1926) and by Rasch (1961):

> It should be possible to omit several test questions at different levels of the scale without affecting the individual's score [measure].

> ... a comparison between two individuals should be
> independent of which stimuli [test questions] within the
> class considered were instrumental for comparison; and
> it should also be independent of which other individuals
> were also compared, on the same or some other occasion.

Taken to the extreme, we can imagine a group of language test takers (reading, writing, speaking, or listening) being invariantly located on a scale without sharing a single item in common (no item is taken by more than one person). This context defines the limit case of omitting items and making comparisons independent of the particular questions answered by any test taker.

More formally, we can contrast a fully crossed p x i design (persons crossed with items) in which all persons take the same set of items with a nested item by person design (i : p) in which all items are unique to each specific person. The more common design in language research is p x i simply because there is no method of data analysis that can extract invariant comparisons from an i : p design unless item calibrations are available from a previous calibration study or are theoretically specified.

But the *i : p* design is routinely encountered in physical science measurement contexts and in health care when, for example, parents report their child's temperature to a pediatrician. Children in different families do not share the same thermometers. Furthermore, the thermometers need not even share the same underlying technology (mercury in a tube versus NexTemp technology, see note 1). Yet there is little doubt that the children can be invariantly ordered and spaced on any of several temperature scales.

The major difference between the typical language testing and temperature taking scenarios is that the same construct theory, engineering specifications and manufacturing quality control procedures have been enforced for each and every thermometer, even though the measurement mechanisms may vary. In addition, considerable resources have been expended in ensuring the measuring unit (°F or °C) has been consistently mapped to the measurement outcome (e.g., column height of mercury or cavity count turning black on a NexTemp thermometer) (Hunter 1980; Latour 1987). Substantive theory, engineering specifications, and functioning metrological networks, not data, render comparable measurement from these disparate thermometers. This contrast illustrates the dominant distinguishing feature between measurement in the physical and educational sciences, including EFL, ESL, and ENL language testing. Educational measurement does not, as a rule, make effective use of substantive theory in the ways the physical sciences do (Taagepera 2008). Nor does educational science embrace metric unification even when constructs (e.g., reading ability) repeatedly assert their separate independent existences (W. Fisher 1997, 1999, 2000; Fisher, Harvey, and Kilgore 1995).

Typical applications of Rasch models in language testing are thin on substantive theory. Rarely is there an a priori specification of the item calibrations (i.e., constrained model). Instead the researcher estimates both person parameters and item parameters from the same $p \times i$ data set. For Kuhn (1961), this practice is at odds with the scientific function of measurement in that substantive theory will almost never be revealed by measuring. Rather "the scientist often seems to be struggling with facts [measurement outcomes, raw scores], trying to force them to conformity with a theory s(he) does not doubt" (163). Kuhn is speaking about substantive construct theory, not axiomatic measurement theory. Demonstrating data fit to a descriptive Rasch model or sculpting a data set by simply eliminating misfitting items and persons and then rerunning the Rasch analysis to achieve satisfactory fit is specifically not the "struggling" Kuhn is referring to.

The gold-standard demonstration that a construct is well specified is the capability to manufacture strictly parallel instruments. A strictly parallel instrument is one in which the correspondence table linking attribute measure to measurement outcome (count correct) is identical although items are different on each parallel instrument. Imagine two four-thousand-word 1300L articles, one on atomic theory and one on mythology. Both articles are submitted to a machine that builds forty-five four-choice cloze items distributed about one item for every eighty to one hundred words. These one-off items are assumed to have calibrations sampled from a normal distribution with a mean equal to 1300L and a standard deviation equal to 132L. With this information, an ensemble Rasch model (Lattanzio, Burdick, and Stenner 2012) can produce a correspondence table linking count correct to Lexile measure. Since the specifications (test length, text measure, text length, and item spread) are identical for the two articles, the correspondence tables will also be identical; on both forms, twenty-five correct answers converts to 1151L, forty correct answers converts to 1513L, and so on. The same basic structure plays out with NexTemp thermometers. A NexTemp thermometer has forty-five cavities. Twenty-five cavities turning black converts to a temperature of 101°F, whereas forty cavities turning black converts to 104°F. In each case theory, engineering specifications and manufacturing guidelines combine to produce strictly parallel instruments for measuring, and in each case it is possible to manufacture large quantities of identical instruments. The capacity to manufacture "strictly" parallel instruments is a milestone in an evolving understanding of an attribute and its measurement. Richard Feynman wrote: "What I cannot create, I do not understand." We demonstrate our understanding of how an instrument works by creating copies that function like the original.

Descriptive Rasch Models versus Causal Rasch Models

Andrich (2004) makes the case that Rasch models are powerful tools precisely because they are prescriptive, not descriptive, and when model prescriptions meet data, anomalies arise. Rasch models invert the traditional statistical data-model relationship by stating a set of requirements that data must meet if those data are to be useful in making measurements. These model requirements are independent of the data. It does not matter if the data are bar presses, counts correct on a reading test, or wine taste preferences. If these data are to be useful in making measures of rat perseverance, reading ability, or vintage quality, all three sets of data must conform to the same invariance requirements. When data fail to fit a model, Rasch measurement theory does not respond by relaxing the invariance requirements and adding, say, an item specific discrimination parameter to improve fit (Rasch 1960; Andrich 1988 2010; Wright 1977, 1999), as does item response theory (Hambleton, Swaminathan, and Rogers 1991). Rather, the Rasch approach is to examine the items serving as the medium for making observations and to change them in ways likely to produce new data conforming to the modeled expectations.

A causal Rasch model (in which item calibrations come from theory, not data) is then doubly prescriptive (Stenner, Stone, and Burdick 2009). First, in accord with Rasch, it is prescriptive regarding the data structures that must be present:

> The comparison between two stimuli should be independent of which particular individuals were instrumental for the comparison; and it should also be independent of which other stimuli within the considered class were or might also have been compared. Symmetrically, a comparison between two individuals should be independent of which particular stimuli within the class considered were instrumental for comparison; and it should also be independent of which other

individuals were also compared, on the same or on some other occasion. (Rasch 1961)

Second, causal Rasch models (Burdick, Stone, and Stenner 2006; Stenner, Burdick, and Stone 2008) prescribe the values imposed by substantive theory on the item calibration estimates. Thus, the data, to be useful in making measures, must conform to both Rasch model invariance requirements *and* to substantive theory invariance requirements as specified by the theoretical item calibrations.

When data meet both sets of requirements, then those data are useful not just for making measures of some vaguely defined construct but are useful for making measures of that precise construct specified by the equation that produced the theoretical item calibrations. We note again that these dual invariance requirements come into stark relief in the extreme case of no connectivity across stimuli or examinees ($i : p$). For example, how are two readers to be measured on the same scale if they share no common text passages or items? If you read *The Hunger Games* and I read *Lord of the Rings* and we each answer distinct sets of machine-generated questions, how would it be possible to realize an invariant comparison of our reading abilities except by means of predictive theory? How else would it be possible to know that you read 250L better than I read, or that you comprehended 95 percent of what you read, whereas I comprehended 75 percent of what I read? Most importantly, by what other means than theory would it be possible to reproduce this result to within a small range of error using another two completely different books as the basis of comparison?

Given that seemingly nothing is in common between the two reading experiences above, invariant comparisons might be thought to be impossible. Yet, as shown in the thermometer example, it is a routine experience for different instruments to be interpreted as informing comparable measures of temperature. Why are we so quick to accept that you have a high-grade fever of 104°F and I have a low-grade fever of 100°F, based on measurements from two different thermometers, and yet find the book reading example inexplicable? The answer lies in well-developed construct theory, rigorously established instrument engineering principles, and uniform metrological conventions (W. Fisher 2009).

Clearly, each of us has had ample confirmation that weight denominated in pounds and kilograms can be accurately measured by any reputable manufacturer's bathroom scale. Experience with diverse bathroom scales has convinced us that, within a pound or two of error, these instruments will produce not just invariant relative differences between two persons but will also meet the more stringent expectation of invariant absolute magnitudes for each individual independent of instrument. Over centuries, instrument engineering has steadily improved to the point that for most purposes, "uncertainty of measurement" (usually interpreted as the standard deviation of a distribution of imagined or actual replications taken on a single person) can be effectively ignored for most bathroom scale applications. And, quite importantly, by convention (i.e., the written or unwritten practice of a community), weight is denominated in standardized units (kilograms or pounds). The choice of any given unit is arbitrary, but what is decisive is that a unit is agreed to by the community and maintained through consistent implementation, instrument manufacture, and reporting. At present, language ability (reading, writing, speaking, and listening) does not enjoy a common construct definition, widely promulgated set of instrument specifications, or conventionally accepted unit of measurement. The challenges that must be addressed in defining constructs, specifying instrument characteristics, and standardizing units include cultural assumptions about number and objectivity, political challenges in shaping legislation, resource allocation, and the expectations and procedures of social scientists (W. Fisher 2012a; W. Fisher and Stenner 2009). In this context, the Lexile Framework for Reading (Stenner, H. Burdick, Sanford, and D. Burdick 2006) stands as an exemplar of how psychosocial measurement can be unified in a manner precisely parallel to the way unification was achieved for length, temperature, weight, and dozens of other useful attributes (Stenner and Stone 2010).

A causal (constrained) Rasch model (Stenner, Stone, and Burdick 2009) that fuses a substantive theory to a set of axioms for conjoint additive measurement affords a much richer context for the identification and interpretation of anomalies than a descriptive (i.e., unconstrained) Rasch model. First, with the measurement model and the substantive theory fixed, anomalies are understood as problems with the data. Attending to the data ideally leads to improved observation models (e.g., new task types) that reduce unintended dependencies and variability. An example of this kind of improvement in measurement was realized when the duke of Tuscany put a top on some of the early thermometers, thus reducing the contaminating influences of barometric pressure on the measurement of temperature. In contrast with the descriptive paradigm dominating much of education science, the duke did not propose parameterizing barometric pressure in the model in the hope that the boiling point of water at sea level, as measured by open top thermoscopes, would then match the model expectations at three thousand feet above sea level.

Second, with both model and construct theory fixed, our task is to produce measurement outcomes that fit the invariance requirements of both measurement theory and construct theory. By analogy, not all fluids are ideal as thermometric fluids. Water, for example, is nonmonotonic in its expansion with increasing temperature. Mercury, in contrast, has many useful properties as a thermometric fluid. Does the discovery that not all fluids are useful thermometric fluids invalidate the concept of temperature? No! In fact, a single fluid with the necessary properties would suffice to validate temperature as a useful construct; hence, the ubiquity of the mercury thermometer. The existence of a persistent invariant framework makes it possible to identify anomalous behavior (water's strange behavior) and interpret it in an expanded theoretical framework (Chang [2004] 2007).

Analogously, finding that not all reading item types conform to the dual invariance requirements of a Rasch model and the Lexile theory does not invalidate either the axioms of conjoint measurement theory or the Lexile reading theory. Rather, the anomalous behaviors of some kinds of text (recipes, calendars, poems) are open invitations to expand the theory to account for these deviations from expectation. Notice here the subtle shift in perspective. We do not need to find one thousand unicorns; one will do to establish the reality of the class. The finding that reader behavior on a minimum of two types of reading tasks can be regularized by the joint actions of the Lexile theory and a Rasch model is sufficient evidence for the reality of the reading construct. Of course, actualizing this scientific reality to make the reading construct a universally uniform and available object in the world will require the investment of significant social, legal, and economic resources (W. Fisher 2009; W. Fisher and Stenner, 2012; W. Fisher 2000a; W. Fisher 2005; W. Fisher 2011).

Equation (1) is a causal Rasch model for dichotomous data, which sets a measurement outcome (expected score) equal to a sum of modeled probabilities.

$$Expected\ score =: \sum_i \frac{e^{(b-di)}}{1 + e^{(b-di)}} \tag{1}$$

The measurement outcome is the dependent variable and the measure (e.g., person parameter, b) and instrument (e.g., the parameters d_i pertaining to the difficulty d of item i) are independent variables. The measurement outcome (e.g., count correct on a reading test) is observed, whereas the measure and instrument parameters are not observed but can be estimated from the response data and substantive theory, respectively. When an interpretation invoking a predictive mechanism is imposed on the equation, the right-side variables are presumed to characterize the process that generates the measurement outcome on the left side. The symbol =: was proposed by Euler circa 1734 to distinguish an algebraic identity from a causal identity (right-hand side causes the left-hand side). This symbol (=:) was reintroduced by Judea Pearl and can be read as indicating that manipulation of the right-hand side via experimental intervention will cause the prescribed change in the left-hand side of the equation. But simple use of an equality (=) does not signal a causal interpretation of the equation.

A Rasch model combined with a substantive theory embodied in a specification equation provides a more or less complete explanation of how a measurement instrument works (Stenner, Stone, and Burdick 2009). A Rasch model in the absence of a specified measurement mechanism is merely a probability model. A probability model absent a theory may be useful for describing or summarizing a body of data, and for predicting the left side of the equation from the right side, but a Rasch model in which instrument calibrations come from a substantive theory that specifies how the instrument works is a causal model. That is, it enables prediction after intervention.

Below we summarize two key distinguishing features of causal Rasch models and highlight how these features can contribute to improved ENL, EFL, and ESL measurement.

1. First, causal Rasch models are individually centered, meaning that a person's measure is estimated without recourse to data on other individuals. The measurement mechanism that transmits variation in the language attribute (within person over time) to the measurement outcome (count correct on a reading test) is hypothesized to function the same way for every person. This hypothesis is testable at the individual level using Rasch model fit statistics.

2. Figuring prominently in the measurement mechanism for language measurement is text complexity. The specification equation used to measure text complexity is hypothesized to function the same way for most text genres and for readers who are ENL, EFL, and ESL. This hypothesis is also testable at the individual level, but aggregations can be made to examine invariance over text types and reader characteristics.

Point one emphasizes that individual measurement is desired and should be the major focus of all assessment. Without this emphasis on the individual, there can be no assessment of growth or progress and no assemblage of group findings. Point two emphasizes the need for a substantive consideration in all measurement inquiries, without which assessment cannot proceed.

EdSphere Reader App

The data for computing empirical text complexity measures came from the reader appliance in EdSphere. Students access tens of millions of professionally authored digital texts in the EdSphere Reader App. Digital articles are drawn from hundreds of periodicals including *Highlights for Children, Boys' Life, Girls' Life, Sports Illustrated, Newsweek, Discovery, Science, The Economist, Scientific American,* and more. Such a large repository of high-quality informational text is required to immerse students with widely varying reading abilities in daily deliberate practice across the K–16 education experience.

Students use three search strategies to locate articles targeted at their Lexile level: (1) click on suggested topics, (2) click the icon "Surprise Me," or the most frequently used method (3) type search terms into "Find a Book or Article" (see figure 4.1). In the example below, a 1069L reader typed "climate change" in the search box and found 13,304 articles close to her reading level. The first text is an 1100L four-page article from *Scientific American* with a short abstract. Readers browse the abstracts and refine the search terms until they find an appropriate length article about their interest topic (or a teacher-assigned topic) at their reading level.

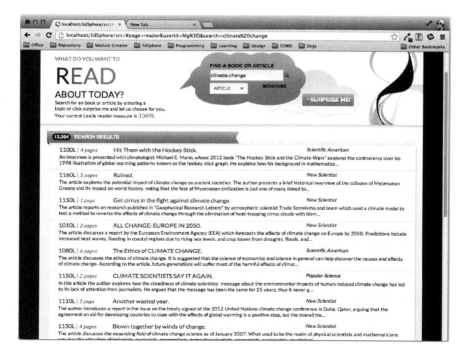

Figure 4.1. Results from student's keyword search for articles about climate change; results include publication titles, publication dates, and (or) page length

Within one second of selecting an article, the machine builds a set of embedded auto-semantic cloze items. In the example below, a student selected the article "A Low-Carbon Future Starts Here" (see figure 4.1). The auto-generated semantic cloze engine created 19 cloze items for this article. Students choose from the four options that appear at the bottom of the page. The incorrect options have similar difficulty and part of speech to the correct answer. The answer is autoscored, and the correct answer is immediately restored in the text and color coded as to whether the student answered correctly or incorrectly. The goal is to read-to-learn, so the system will not let an incorrect choice undermine comprehension.

Three instructional supports are built into the Reader App to facilitate comprehension. First, suggested strategies are presented to students during the reading process. Second, students have access to an in-line dictionary and thesaurus (one-click access). Finally, a text-to-speech engine has been integrated into EdSphere, allowing words, phrases, or sentences to be machine-spoken to the reader.

Text Complexity 719s with Artifact Correction

Figure 4.2 presents the results of a multiyear study of the relationship between theoretical text complexity as measured by the Lexile Analyzer (freely available for noncommercial use at Lexile.com) and empirical text complexity as measured by the EdSphere platform. Each of the 719 articles included in this study was evaluated by the analyzer for semantic demand (log transformed frequency of each word's appearance in a multibillion-word corpus) and syntactic demand (log transformed mean sentence length). While the text preprocessing (defining what constitutes a word) involves thousands of lines of code, modern computing enables the measurement of the Bible or Koran in several seconds.

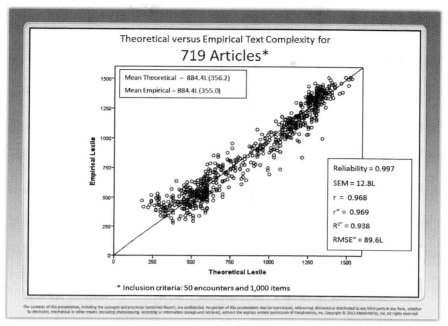

Figure 4.2

The EdSphere platform enables students to select articles of their choosing from a collection of over one hundred million articles that have been published and measured over the past twenty years. As a student's reading ability grows, a 200L window moves up the scale (100L below the student's ability to 100L above), and all articles relevant to a reader's search term that have text complexity measures in the window are returned to the reader. The machine generates a four-choice cloze every seventy-to eighty words, and the count correct combined with the readers Lexile measure is used to compute an empirical text complexity for the article averaged over at least fifty readers and at least one thousand items.

The 719 articles chosen for this study were the first articles to meet the dual requirements of at least fifty readers and at least one thousand item responses. Accurately estimated reader measures were available prior to the encounter between an article and a reader. Thus, each of the articles has a theoretical text complexity measure from the Lexile Analyzer and an empirical text complexity from EdSphere. The correlation between theory and empirical text complexity is r = 0.968 (r^2 = .938).

Connecting the Psychometrics to an Instructional
Theory for Language Development

Work on connecting causal Rasch models to theories of language development (Hanlon 2013; Swartz, Hanlon, Stenner, and Childress 2015) has made extensive use of Ericsson's theory of deliberate practice in the acquisition of language expertise. Deliberate practice is a core tenant of Ericsson's theory of expertise development. Hanlon (2013) distills five core principles of deliberate practice in the development of reading ability: (1) targeted practice reading text that is not too easy and not too hard, (2) real-time corrective feedback on embedded response requirements, (3) distributed practice over a long period of time (years, decades), (4) intensive practice that avoids burnout, and (5) self-directed options when one-on-one coaching is not available. Each of these principles, when embedded into instructional technologies, benefits from individually centered psychometric models in which, for example, readers and text are measured in a common unit.

Swartz, Hanlon, Stenner, and Childress (2015) provide a complete description of EdSphere, including its history and components. The EdSphere technology is designed to immerse students in deliberate practice in reading, writing, and content vocabulary, as well as with conventions of standard English: "These principles of deliberate practice are strengthened by four choice psychometrically sound assessment approaches into learning activities. For example students respond to cloze items while reading, compose short and long constructed responses in response to prompts, correct different kinds of convention errors (i.e. spelling, grammar, punctuation, capitalization) in authentic text, and select words with common meanings from a thesaurus-based activity. Each item encountered by students is auto-generated and auto-scored by software. The results of these learning embedded assessments are especially beneficial when assessment item types are linked to a developmental scale" (Swartz, Hanlon, Stenner, and Childress 2015).

Figure 4.3 is an individual-centered reading growth trajectory denominated in Lexiles. All data comes from EdSphere. Student 1528 is an ESL seventh-grade male (first language Spanish) who read 347 articles of his choosing (138,695 words) between May 2007 and April 2011. Each solid dot corresponds to a monthly average Lexile measure. The growth curve fits the monthly means quite well, and this young man is forecasted (big dot on the far right of the figure) to be a college-ready reader when he graduates from high school in 2016. The open dots distributed around zero on the horizontal axis are the expected performance minus observed performance (in percent) for each month. Expected performance is computed using the Rasch model, inputs for each article's text complexity, and the updated reader's ability measure. Given these inputs, EdSphere forecasts a percent correct for each article encounter. The observed performance is the observed percentage correct for the month. The difference between what the substantive theory (Lexile Reading Framework) expects in cooperation with the Rasch model and what is actually observed is plotted by month. The upper left-hand corner of the graphic summarizes the expected percentage correct over the four years (73.5 percent) and observed percentage correct (71.7 percent) across the 3,342 items taken by this reader. Note that EdSphere is dynamically matching text complexity of the articles the reader can choose to the growing reader ability over time. So this graphic describes a within-person (intraindividual) test of the quantitative hypothesis: can EdSphere trade-off a change in reader ability for a change in text complexity to hold constant the success rate (comprehension)? For this reader, the answer appears to be a resounding yes!

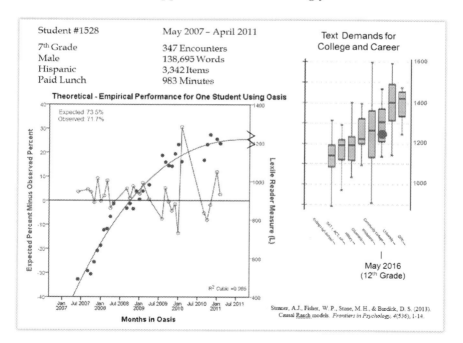

Figure 4.4 is a graphical depiction of the 99 percent confidence interval for the artifact-corrected correlation between theoretical and empirical text complexity. The artifacts included measurement error, double range restriction, and construct invalidity. The artifact-corrected

correlation (coefficient of theoretical equivalence) is slightly higher than r = 1.0, suggesting that the Lexile theory accounts for all of the true score variation in the empirical text complexity measures. The reader may be puzzled about how a correlation can be higher than r = 1.0; of course it can't be, but an artifact corrected correlation can be if the artifactors used in the correlation are, perhaps due to a sampling error, lower than their population values.

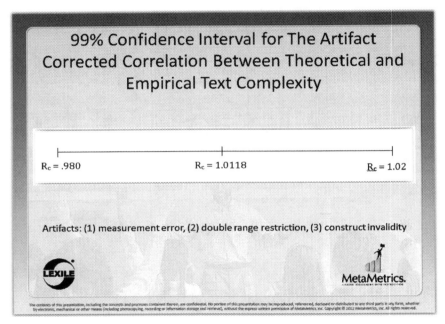

Figure 4.4

Notes

1. The NexTemp thermometer is a small plastic strip pocked with multiple enclosed cavities. In the Fahrenheit version, forty-five cavities arranged in a double matrix serve as the functioning end of the unit. Spaced at 0.2°F intervals, the cavities cover a range from 96.0°F to 104.8°F. Each cavity contains three cholesteric liquid crystal compounds and a soluble additive. Together, this chemical composition provides discrete and repeatable change-of-state temperatures consistent with the device's numeric indicators.

Change of state is displayed optically (cavities turn from green to black) and is easily read.

2. Text complexity is predicted from a construct specification equation incorporating sentence length and word frequency components. The squared correlation of observed and predicted item calibrations across hundreds of tests and millions of students over the last fifteen years averages about .93. Recently available technology for measuring reading ability employs computer-generated items built on the fly for any continuous prose text in a manner similar to that described for mathematics items by Bejar and colleagues (Bejar, Lawless, Morley, Wagner, Bennett, and Revuelta 2003). Counts correct are converted into Lexile measures via a Rasch model estimation algorithm employing theory-based calibrations. The Lexile measure of the target text and the expected spread of the cloze items are given by theory and associated equations. Differences between two readers' measures can be traded off for a difference in Lexile text measures. When the item generation protocol is uniformly applied, the only active ingredient in the measurement mechanism is the choice of text complexity (choosing a 500L article) and the cloze protocol implemented by the machine.

In the temperature example, a uniform increase or decrease in the amount of soluble additive in each cavity changes the correspondence table that links the number of cavities that turn black to degrees Fahrenheit. Similarly, an increase or decrease in the text demand (Lexile) of the passages used to build reading tests predictably alters the correspondence table that links count correct to Lexile reader measure. In the former case, a temperature theory that works in cooperation with a Guttman model produces temperature measures. In the latter case, a reading theory that works in cooperation with a Rasch model produces reader measures. In both cases, the measurement mechanism is well understood, and we exploit this understanding to address a vast array of counterfactuals (Woodward 2003). If things had been different (with the instrument or object of measurement), we could still answer the question as to what then would have happened to what we observe (i.e., the measurement outcome). It is this kind of relation that illustrates the meaning of the expression "There is nothing so practical as a good theory" (Lewin 1951).

Box 12: From Cloud to Arrow

This figure pictures a simple analogy (just an illustration, not intended to be carried too far) to represent the process of many scientific investigations. Scientific investigations often begin from an amorphous state, progressing only when some direction, goal, or strategy is pursued (substantive theory supporting a causal hypothesis). Only when a conjectured plan has been clearly conceptualized does progress occur and clear-cut strategies emerge. Such approaches, once conceived, may be successful, or they may be dead wrong. The critical point is to have a strategy clearly delineated so that its causality is clearly stated and its hypotheses open to critical evaluation. Karl Popper encourages one to speculate but especially to critically test via refutation. This approach is warranted, but we make slight modifications to this approach by asserting the importance of substantive theory rather than simply stating conjectures. While we recognize that conjectures may be and have been profitable, this approach has not been normative. Instead, most breakthroughs occur by wise and rational processes.

The arrow symbolizes a directive approach to causal models. This approach requires delineating the model. We have found the triplet approach a good paradigm to follow in specifying the causal model of Rasch, showing *outcome* = *d* * *b* together with clear and distinct characteristics for what *b* and *d* are expected to represent. This is not to say that a model must have only two independent variables but only that the specification equation should be succinct (refer to box 13, which delineates specification equation details).

Box 13: Specification Equations in Causal Models

A specification equation delineates the essential characteristics that define the dependent and independent variables. It is typically composed of fewer rather than more characteristics, not because life is simple but because we cannot conceptualize the complexity of life and nature without becoming completely overwhelmed. We must determine the essential characteristics, as physicists do when dealing with their domain. We illustrate these matters with several examples. An equation is specific, but a clear explication of the variables is equally important.

1. **Windchill.** Reporting windchill is especially popular now but has always been appreciated by those who can clearly tell the difference between how they feel in cold weather with a wind versus cold weather without wind. As the air temperature falls, the wind increases the chilling effect. The original formula specified the windchill index as kcal/m²/h using wind velocity, m/s, and air temperature in °C (WCI = 10{√V} − Vt + 10). Successive formulae have become much more detailed, but the variables still consist of essentially wind and temperature predicting windchill.

2. **KCT.** Stenner and Stone conducted independent investigations and analyses using different KCT data and found the conceptual specifications of distance, reverses in tapping, and overall length of the tapping series to be the essential characteristics of the instrument for determining individual success or failure. These three characteristics comprised the theoretical causal order. Specification equations validated the order and confirmed the substantive theory.

Box 14: The Concept of a Measurement Mechanism

"And in technology, as well as in basic science, to
explain a fact is to exhibit the mechanism(s) that makes
the system in question tick." (Bunge 2004, 182)

In 1557, the Welshman Robert Recorde remarked that no two things could be more alike (i.e., more equivalent) than parallel lines, and thus was born the equal sign. Equation (1) shows the familiar Rasch model for dichotomous data, which sets a measurement outcome (raw score) equal to a sum of modeled probabilities. The measurement outcome is the dependent variable and the measure (e.g., person parameter, b) and instrument (e.g., item parameters di's) are independent variables. The measurement outcome (e.g., count correct on a reading test) is observed, whereas the measure and instrument parameters are not observed but can be estimated from the response data. When a mechanismic interpretation[1] is imposed on the equation, the right-hand side (r.h.s.) variables are presumed to characterize the process that generates the measurement outcome on the left-hand side (l.h.s.). An illustration of how such a mechanism can be exploited is given in Stone (2002b). The item map for the Knox Cube Test analysis had a logit gap of l. The specification equation was used to build an item that theory asserted would fill in the gap. Subsequent data analysis confirmed the theoretical prediction of the Rasch relationship:

$$Expected\ raw\ score =: \sum_i \frac{e^{(b-di)}}{1 + e^{(b-di)}} \qquad (1)$$

Typically, the item calibrations (di's) are assumed to be known, and the measure parameter is iterated until the equality is realized (i.e., the sum of the modeled probabilities equals the measurement outcome). How is this equality to be interpreted? Are we only interested in the algebra or is something more happening?

Freedman (1997) proposed three uses for a regression equation like the one above:

1.1) To describe or summarize a body of data
1.2) To predict the *l.h.s.* from the *r.h.s.*
1.3) To predict the *l.h.s.* after manipulation or intervention on one or more *r.h.s.* variables (measure parameter and/or instrument parameters)

Description and summarization possess a reducing property in that they abstract away incidentals to focus on what matters in a given context. In a rectangular persons-by-items data matrix (with no missing data), there are *np x ni* observations. Equations like those above summarize the data using only *np + ni - 1* independent parameters. Description and summarization are local in focus. The relevant concept is the extant data matrix with no attempt to answer questions that might arise in the application realm concerning "what if things were different." Note that if interest centers only on the description and summary of a specific data set, additional parameters can be added, as necessary, to account for the data.

Prediction typically implies the use of the extant data to project into an as yet unobserved context/future in the application realm. For example, items from the extant data are used to compute a measure for a new person, or person parameters are used to predict how these persons will perform on a new set of items. Predictions like these rest on a set of claims of invariance. New items and new persons are assumed to behave as previous persons and items behaved in the extant data set. Rasch-fit statistics (for persons and items) are available to test for certain violations of these assumptions of invariance (Smith 2000).

Rasch models are probabilistic models that are fundamentally associational and thus cannot and do not, alone, support a causal interpretation of equation (1) (Woodward 2003). Note that equation (1) can support a predictive interpretation if the equality is taken to satisfy a simple if-then condition. A causal interpretation of equation (1) requires successful predictions under manipulation of the measure parameter, the instrument parameters, or ideally, under conjoint manipulation of the two parameters. Conjoint manipulation up and down the scale directly tests for the trade-off property that holds only when the axioms of additive conjoint measurement are satisfied (Kyngdon 2008a).

To explain how an instrument works is to detail how it generates the count it produces (measurement outcome) and what characteristics of the measurement procedure affect that count. This kind of explanation is neither just statistical nor synonymous with prediction. Instead, the explanation entails prediction under intervention: if I wiggle this part of the mechanism, the measurement outcome will be different by this amount. As noted by Hedström (2005), "Theories based on fictitious assumptions, even if they predict well, give incorrect answers to the question of why we observe what we observe" (108). Rasch models, absent a substantive theory capable of producing theory-based instrument calibrations, may predict how an instrument will perform with another subject sample (invariance) but can offer only speculation in answer to the question "How does this instrument work?" Rasch models without theory are not predictive under intervention and thus are not causal models.

Measurement mechanism is the name given to just those manipulable features of the instrument that cause invariant measurement outcomes for objects of measurement that possess identical measures. A measurement mechanism explains by opening the black box and showing the cogs and wheels of the instrument's internal machinery. A measurement mechanism provides a continuous and contiguous chain of causal links between the encounter of the object of measurement and instrument and the resulting measurement outcome (Elster 1989). We say that the measurement outcome (e.g., raw score) is explained by explicating the mechanism by which the measurement outcome is brought about. In this view, to respond with a recitation of the Rasch equation for converting counts into measures, to reference a person by item map, to describe the directions given to the test taker, to describe an item-writing protocol, or simply to repeat the construct label more slowly and loudly (e.g., extroversion), provide nonanswers to the question "How does this instrument work?"

Although the sociologist Peter Hedström (2005) was concerned with the improvement of macro theory, several of his reasons for favoring mechanistic explanations apply to measurement science in general:

2.1) Detailed specifications of mechanisms result in more intelligible explanations.

2.2) A focus on mechanisms rather than, for example, item types, reduces theoretical fragmentation by encouraging consideration of the possibility that many seemingly distinct instruments (e.g., reading tests) with different item types and construct labels may in fact share a common measurement mechanism.

2.3) The requirement for mechanistic explanations helps to eliminate spurious causal accounts of how instruments work.

Measurement mechanisms as theoretical claims make point predictions under intervention: when we change (via manipulation or intervention) either the object measure (e.g., reader experiences growth over a year) or measurement mechanism (e.g., increase text complexity by 200L). The mechanismic narrative and associated equations enable a point prediction on the consequent change in the measurement outcome (i.e., count correct). Notice how this process is crucially different from the prediction of the change in the measurement outcome based on the selection of another previously calibrated instrument with known instrument calibrations. Selection is not intervention in the sense used here. Our sampling from banks of previously calibrated items is likely to be completely atheoretical, relying, as it does, on empirically calibrated items/instruments. In contrast, if we modify the measurement mechanism (manipulate variables in the specification to generate new items) rather than select previously calibrated measurement mechanisms, we must have intimate knowledge of how the instrument works. Atheoretical psychometrics is characterized by the aphorism "test the predictions, never the postulates" (Jasso 1988, 4), whereas theory-referenced measurement, with its emphasis on measurement mechanisms, says to test the postulates, rather than the predictions. Those who fail to appreciate this distinction will confuse invariant predictors with genuine causes of measurement outcomes.

A Rasch model combined with a substantive theory embodied in a specification equation provides a more or less complete explanation of how a measurement instrument works (Stenner, Smith, and Burdick 1983). A Rasch model in the absence of a specified measurement mechanism is merely a probability model; a probability model absent a theory may be useful for (1.1) and (1.2), whereas a Rasch model in which instrument calibrations come from a substantive theory that specifies how the instrument works is a causal model. That is, it enables prediction after intervention (1.3):

Causal models (assuming they are valid) are much more informative than probability models: A joint distribution tells us how probable events are and how probabilities would change with subsequent observations, but a causal model also tells us how these probabilities would change as a result of external interventions ... Such changes cannot be deduced from a joint distribution, even if fully specified. (Pearl [2000] 2009, 22)

A mechanismic narrative provides a satisfying answer to the question of how an instrument works. Below are two such narratives for a thermometer designed to take human temperature (3.1) and a reading test (3.2).

3.1) The Nextemp thermometer is a thin, flexible, paddle-shaped plastic strip containing multiple cavities. In the Fahrenheit version, the 45 cavities are arranged in a double matrix at the functioning end of the unit. The columns are spaced at 0.2°F intervals covering the range of 96.0°F to 104.8°F. . . . Each cavity contains a chemical composition comprised of three cholesteric liquid crystal compounds and a varying concentration of a soluble additive. These chemical compositions have discrete and repeatable change-of-state temperatures consistent with an empirically established formula to produce a series of change-of-state temperatures consistent with the indicated temperature points on the device. The chemicals are fully encapsulated by a clear polymeric film, which allows observation of the physical change but prevents any user contact with the chemicals. When the thermometer is placed in an environment within its measure range, such as 98.6°F (37.0°C), the chemicals in all of the cavities up to and including 98.6°F (37.0°C) change from a liquid crystal to an isotropic clear liquid state. This change of state is accompanied by an optical change that is easily viewed by a user. The green component of white light is reflected from the liquid crystal state but is transmitted through the isotropic liquid state and absorbed by the black background. As a result, those cavities containing compositions with threshold temperatures up to and including 98.6°F (37.0°C) appear black, whereas those with transition temperatures of 98.6°F (37.0°C) and higher continue to appear green (Medical Indicators 2006, 1–2).

3.2) The MRW technology for measuring reading ability employs computer generated four-option multiple choice cloze items "built on-the-fly" for any continuous prose text. Counts correct on these items are converted into Lexile measures via an applicable Rasch model. Individual cloze items are one-off and disposable. An item is used only once. The cloze and foil selection protocol ensures that the correct answer (cloze) and incorrect answers (foils) match the vocabulary demands of the target text. The Lexile measure of the target text and the expected spread of the cloze items are given by a proprietary text theory and associated equations. A difference between two reader measures can be traded off for a difference in Lexile text measures to hold count correct (measurement outcome) constant. Assuming a uniform application of the item generation protocol, the only active ingredient in the measurement mechanism is the choice of text with the requisite semantic (vocabulary) and syntactic demands.

In the first example, if we uniformly increase or decrease the amount of additive in each cavity, we change the correspondence table that links the number of cavities that turn black to a Fahrenheit degree count. Similarly, if we increase or decrease the text complexity (Lexile) of the passages used to build reading tests, we predictably alter the correspondence table that links count correct to Lexile reader measure. In the former case, a temperature theory that works in cooperation with a Guttman model produces temperature measures. In the latter case, a reading theory that works in cooperation with a Rasch model produces reader measures. In both cases, the measurement mechanism is well understood, and we exploit this understanding to answer a vast array of "W" questions (see Woodward 2003): if things had been different (with the instrument or object of measurement), what then would have happened to what we observe (i.e., the measurement outcome)?

To explain a measurement outcome, "One must provide information about the conditions under which [the measurement outcome] would change or be different. It follows that the generalizations that figure in explanations [of measurement outcomes] must be change-relating … Both explainers [e.g., person parameters and item parameters] and what is explained [measurement outcomes] must be capable of change, and such changes must be connected in the right way" (Woodward 2003, 234).

The Rasch model tells us the right way to connect object measures, instrument calibrations, and measurement outcomes. Substantive theory tells us what interventions/changes can be made to the instrument to offset a change to the measure for an object of measurement to hold constant the measurement outcome. Thus, a Rasch model in cooperation with a substantive theory dictates the form and substance of permissible conjoint interventions. A Rasch analysis, absent a construct theory and associated specification equation, is a black box, and "as with any black-box computational procedures, the illusion of understanding is all too easy to generate" (Humphreys 2004, 132).

Footnotes

1. The term mechanismic was coined by Bunge (2004) to emphasize the nonmechanical features of some mechanisms.
2. In applied mathematics, we typically distinguish between the mathematical realm and the application realm.

Box 15: Alfred Binet and Constructive Measurement

Alfred Binet was a pioneer in producing measures of intellectual development, but he inherited a negative reputation from those who borrowed his tests without following his course of inquiry. From his first efforts in 1905, Binet (1908, and subsequent revisions) produced successive editions that incorporated improvements to "measure the intellectual development of children whose ages are between three and twelve years [...] of age, or is advanced or retarded" (p.8). Changes in terminology should not impede our understanding of what he was trying to accomplish. Changes in terminology should not impede our understanding of what he was trying to accomplish" (8).

Binet proposed a concept of intelligence whereby,

> The general formula is that an individual is normal when he can conduct himself without having the need of the tutelage of others, when he earns sufficient income for his needs, and finally when his intelligence does not take into work of a lower classification than that of his parents [...]

Retardation is an idea related to a host of circumstances that must be kept in mind in judging each particular case. The judgement must be made on a synthesis of resultants. (88)

He advocated a new research approach called the *psychogenetic method*, a position not unlike what developmental psychologists now designate as epigenetic development, a synthesis of inherited human attributes expressed within an environmental context, discriminating structure from function.

Binet spoke of mental level ("niveau mental") and not mental age in an approach to measuring intelligence in general. He was critical of Spearman's approach to a general level of intelligence, and also of those promoting special abilities assessment.

His approach to normal thought processes supported three operations of the mind: direction, correction/criticism, and adaptation. Intelligence acted within a defined range of possibilities from which to choose, using that capacity to self-correct and change, and by adapting or being open to new circumstances. His scales sought to order individuals by utilizing his findings on what tests ("test" was used instead of "item," a wise decision) defined each age level, using 50 percent as the discriminating mark of a useful test. Passing the test placed the person in the age level. Missing the item did not. Five examples at each age were utilized to reduce the effect of chance. Binet never let his data divert him from his goal. In fact, he came to identify two types of texts that he deemed especially utilitarian for identifying intellectual ability—comprehension and absurdities. The former is illustrated by "Before taking part in an important matter, what should you do?" and the latter, "There was an accident yesterday, but it wasn't serious; only forty-eight people were killed." These two types of tests were to be designed to be age appropriate.

Binet's work is now more than one hundred years old. His approach was lost by over-attention to the production of new editions made by researchers operating without a substantive theory to guide their investigations, completely ignoring Binet's three key principles of intellectual development—direction, correction/criticism, and adaptation. Binet grasped the causal expression of a specification equation, and had identified some key tests to tap these three elements. This insight was not recognized by subsequent investigators who failed to grasp the importance of his substantive theory, focusing only upon writing helter-skelter items analyzed with dubious procedures.

5

CAUSATION AND MANIPULABILITY

Manipulability theories of causation, according to which causes are to be regarded as handles or devices for manipulating effects, have considerable intuitive appeal and are popular among social scientists and statisticians. This chapter surveys several prominent versions of such theories advocated by philosophers, and the many difficulties they face. Philosophical statements of the manipulationist approach are generally reductionist in aspiration and assign a central role to human action. These contrast with recent discussions employing a broadly manipulationist framework for understanding causation, including those indebted to the computer scientist Judea Pearl, which are nonreductionist and rely instead on the notion of an intervention. This is simply an appropriately exogenous causal process; it has no essential connection with human action. This interventionist framework manages to avoid at least some of the difficulties faced by traditional philosophical versions of the manipulability theory.

Introduction

A commonsensical idea about causation is that causal relationships are relationships that are potentially exploitable for purposes of manipulation and control: very roughly, if C is genuinely a cause of E, then if I can manipulate C in the right way, this should be a way of manipulating or changing E. This idea is the cornerstone of manipulability theories of causation developed by philosophers such as Gasking (1955), Collingwood (1940), von Wright (1971), Menzies and Price (1993), and Woodward (2003). It is also an idea that is advocated by many nonphilosophers. For example, in their extremely influential text on experimental design Cook and Campbell (1979) write:

> The paradigmatic assertion in causal relationships is that
> manipulation of a cause will result in the manipulation of an

effect ... Causation implies that by varying one factor I can make another vary. (36)

Similar ideas are commonplace in econometrics and in the so-called structural equations or causal modeling literature and very recently have been forcefully reiterated by Judea Pearl in an impressive book-length treatment of causality (Pearl [2000] 2009).

To a large extent, however, recent philosophical discussion has been unsympathetic to manipulability theories: it is claimed both that they are unilluminatingly circular and that they lead to a conception of causation that is unacceptably anthropocentric or at least insufficiently general in the sense that they are linked too closely to the practical possibility of human manipulation (e.g., Hausman 1986, 1998). Both objections seem prima facie plausible. Suppose that X is a variable that takes one of two different values, 0 and 1, depending on whether some event of interest occurs. Then for an event or process M to qualify as a manipulation of X, it would appear that there must be a causal connection between M and X: to manipulate X, one must *cause* it to change in value. How then can we use the notion of manipulation to provide an account of causation? Moreover, it is uncontroversial that causal relationships can exist in circumstances in which manipulation of the cause by human beings is not practically possible—think of the causal relationship between the gravitational attraction of the moon and the motion of the tides or causal relationships in the very early universe. How can a manipulability theory avoid generating a notion of causation that is so closely tied to what humans can do that it is inapplicable to such cases?

As remarked above, the generally negative assessment of manipulability theories among philosophers contrasts sharply with the widespread view among statisticians, theorists of experimental design, and many social and natural scientists that an appreciation of the connection between causation and manipulation can play an important role in clarifying the meaning of causal claims and understanding their distinctive features. This in turn generates a puzzle. Are nonphilosophers simply mistaken in thinking that focusing on the connection between causation and manipulation can tell us something valuable about causation? Does the widespread invocation of something like a manipulability conception among practicing scientists show that the usual philosophical criticisms of manipulability theories of causation are misguided?

The ensuing discussion is organized as follows. Sections 2 and 3 describe two of the best-known philosophical formulations of the manipulability theory—those due to von Wright (1971) and Menzies and

Price (1993)—and explores certain difficulties with them. Section 4 argues that the notion of a free action cannot play the central role it is assigned in traditional versions of manipulability theories. Section 5 introduces the notion of an intervention, which allows for a more adequate statement of the manipulability approach to causation and which has figured prominently in recent discussion. Section 6 considers Pearl's "interventionist" formulation of a manipulability theory and an alternative to it, due to Woodward (2003). Sections 7 and 8 take up the charge that manipulability theories are circular. Section 9 returns to the relationship between interventions and human actions, while section 10 compares manipulability accounts with David Lewis's closely related counterfactual theory of causation. Sections 11, 12, and 13 consider the scope of manipulability accounts, while section 14 considers some recent objections to such accounts.

As we shall see, the difference between assessments of manipulability accounts of causation within and outside of philosophy derive from the different goals or aspirations that underlie the versions of the theory developed by these two groups. Philosophical defenders of the manipulability conception have typically attempted to turn the connection between causation and manipulability into a reductive analysis: their strategy has been to treat as primitive the notion of manipulation (or some related notion like agency, or bringing about an outcome as a result of a free action) to argue that this notion is not itself causal (or at least does not presuppose all the features of causality the investigator is trying to analyze), and then to attempt to use this notion to construct a noncircular reductive definition of what it is for a relationship to be causal. Philosophical critics have (quite reasonably) assessed such approaches in terms of this aspiration (i.e., they have tended to think that manipulability accounts are of interest only insofar as they lead to a noncircular analysis of causal claims) and have found the claim of a successful reduction unconvincing. By contrast, statisticians and other nonphilosophers who have explored the link between causation and manipulation generally have not had reductionist aspirations—instead their interest has been in unpacking what causal claims mean and in showing how they figure in inference by tracing their interconnections with other related concepts (such as manipulation) but without suggesting that the notion of manipulation is itself a causally innocent notion.

It is the impulse toward reduction that generates the other feature that critics have found objectionable in standard formulations of the manipulability theory. To carry through the reduction, one needs to show that the notion of agency is independent of or prior to the notion of causality, and this in turn requires that human actions or manipulations

111

be given a special status—they can't be ordinary causal transactions but must instead be an independent fundamental feature of the world in their own right. This seems problematic both on its own terms (it is prima facie inconsistent with various naturalizing programs) and in that it leads directly to the problem of anthropocentricity: if the only way in which we understand causation is by means of our prior grasp of an independent notion of agency, then it is hard to see what could justify us in extending the notion of causation to circumstances in which manipulation by human beings is not possible and the relevant experience of agency unavailable. As we shall see, both von Wright and Menzies and Price struggle, not entirely successfully, with this difficulty.

The way out of these problems is to follow writers like Pearl in reformulating the manipulability approach in terms of the notion of an intervention, where this is characterized in purely causal terms that make no essential reference to human action. Some human actions will qualify as interventions, but they will do so by virtue of their causal characteristics, not because they are free or carried out by humans. This "interventionist" reformulation allows the manipulability theory to avoid reproducing a number of counterexamples derived from more traditional versions of the theory. Moreover, when so reformulated, the theory may be extended readily to capture causal claims in contexts in which human manipulation is impossible. However, the price of such a reformulation is that we lose the possibility of a reduction of causal claims to claims that are noncausal. Fortunately (or so sections 7 and 8 argue) an interventionist formulation of a manipulability theory may be nontrivial and illuminating even if it fails to be reductive.

An Early Version of an Agency Theory

In an early version of an agency theory, von Wright (1971) describes the basic idea as follows:

> ... to think of a relation between events as causal is to think of it under the aspect of (possible) action. It is therefore true, but at the same time a little misleading to say that if p is a (sufficient) cause of q, then if I could produce p I could bring about q. For *that* p is the cause of q, I have endeavored to say here, *means* that I could bring about q, if I could do (so that) p. (74)

To the objection that "doing" or "producing" is already a causal notion and hence not something to which we can legitimately appeal to elucidate the notion of causation, von Wright responds as follows:

> The connection between an action and its result is intrinsic, logical and not causal (extrinsic). If the result does not materialize, the action simply has not been performed. The result is an essential "part" of the action. It is a *bad* mistake to think of the act(ion) itself as a cause of its result. (67–8)

Here we see a very explicit attempt to rebut the charge that an account of causation based on agency is circular by contending that the relation between an action (or a human manipulation) and its result is not an ordinary causal relation. Moreover, von Wright readily embraces the further conclusion that seems to follow from this: human action must be a concept that, in our understanding of the world, is just as "basic" as the notion of causality (74).

Given the logical structure of von Wright's views, it is also not surprising to find him struggling to make sense of the idea that there can be causal relations involving events that human beings cannot in fact manipulate. He writes:

> The eruption of Vesuvius was the cause of the destruction of Pompeii. Man can through his action destroy cities, but he cannot, we think, make volcanos erupt. Does this not prove that the cause-factor is not distinguished from the effect-factor by being in a certain sense capable of manipulation? The answer is negative. The eruption of a volcano and the destruction of a city are two very complex events. Within each of them a number of events or phases and causal connections between them may be distinguished. For example, that when a stone from high above hits a man on his head, it kills him. Or that the roof of a house will collapse under a given load. Or that a man cannot stand heat above a certain temperature. All these are causal connections with which we are familiar from experience and which are such that the cause-factor typically satisfies the requirement of manipulability. (70)

Von Wright's view is that to understand a causal claim involving a cause that human beings cannot manipulate (e.g., the eruption of a volcano), we must interpret it in terms of claims about causes that human beings *can* manipulate (impacts of falling stones on human heads and so on). We will return to this general idea below in connection with Price and Menzies, but it is worth noting that it faces an obvious problem. If we try to explain what it means to say that different galaxies attract one another gravitationally by contending that such interactions are in some relevant respects similar to gravitational interactions with which we are familiar or have experience (people and projectiles falling to earth), we need to explain what "similar" means and it is very hard to see how to do this within the framework of an agency theory. The relevant notion of similarity does not seem to be a notion that can be spelled out in terms of similarities in human experiences of agency. Either we explain the relevant notion of similarity in straightforwardly causal terms that seem to have nothing to do with agency (e.g., we say that the similarity consists in the fact that the same gravitational force law is operative in both cases), in which case we have effectively abandoned the agency theory, or else we are led to the conclusion that causal claims involving unmanipulable causes like galaxies involve a conception of causality that is fundamentally different from the conception that is applicable to manipulable causes.

A More Recent Version of an Agency Theory

A very similar dialectic is at work in an extremely interesting paper by Peter Menzies and Huw Price (1993) and by Price alone (1991, 1992), which represents the most detailed and sustained attempt in the recent philosophical literature to develop an "agency" theory of causation. Price and Menzies's basic thesis is that:

... an event *A* is a cause of a distinct event *B* just in case bringing about the occurrence of *A* would be an effective means by which a free agent could bring about the occurrence of *B* (1993, 187).

They use this connection between free agency and causation to support a probabilistic analysis of causation (according to which "*A* causes *B*" can be plausibly identified with "*A* raises the probability of *B*") provided that the probabilities appealed to are what they call "agent probabilities," where:

[a]gent probabilities are to be thought of as conditional probabilities, assessed from the agent's perspective under the supposition that an antecedent condition is realized *ab*

114

> *initio*, as a free act of the agent concerned. Thus the agent probability that one should ascribe to B conditional on A is the probability that B would hold were one to choose to realize A. (1993, 190)

Thus the idea is that the agent probability of B conditional on A is the probability that B would have provided that A has a special sort of status or history—in particular, assuming that A is realized by a free act. A will be a cause of B just in case the probability of B conditional on the assumption that A is realized by a free act is greater than the unconditional probability of B; A will be a spurious cause of B just in case these two probabilities are equal. As an illustration, consider a stock example used by philosophers—a structure in which atmospheric pressure, represented by a variable Z, is a common cause of the reading X of a barometer and the occurrence of a storm Y, with no causal relationship between X and Y. X and Y will be correlated, but Price and Menzies's suggestion is that conditional on the realization of X by a free act, this correlation will disappear, indicating that the correlation between X and Y is spurious and does not reflect a causal connection from X to Y. If, by contrast, this correlation were to persist, this would be an indication that X was after all a cause of Y. What "free act" might mean in this context will be explored below, but I take it that what is *intended*—as opposed to what Price and Menzies actually say—is that the manipulation of X should satisfy the conditions we would associate with an ideal experiment designed to determine whether X causes Y—thus, for example, the experimenter should manipulate the position of the barometer dial in a way that is independent of the atmospheric pressure Z, perhaps by setting its value after consulting the output of some randomizing device.

Like von Wright, Price and Menzies attempt to appeal to this notion of agency to provide a noncircular, reductive analysis of causation. They claim that circularity is avoided because we have a grasp of the *experience* of agency that is independent of our grasp of the general notion of causation.

The basic premise is that from an early age, we all have direct experience of acting as agents. That is, we have direct experience not merely of the Humean succession of the events in the external world but of a very special class of such successions: those in which the earlier event is an action of our own, performed in circumstances in which we both desire the later event, and believe that it is more probable given the act in question than it would be otherwise. To put it more simply, we all have direct personal experience of doing one thing and thence achieving another.

... It is this common and commonplace experience that licenses what amounts to an ostensive definition of the notion of 'bringing about'. In other words, these cases provide direct non-linguistic acquaintance with the concept of bringing about an event; acquaintance which does not depend on prior acquisition of any causal notion. An agency theory thus escapes the threat of circularity. (1993, 194–5)

Again, like von Wright, Menzies and Price recognize that once the notion of causation has been tied in this way to our "personal experience of doing one thing and hence achieving another," a problem arises concerning unmanipulable causes. To use their own example, what can it mean to say that "the 1989 San Francisco earthquake was caused by friction between continental plates" (195) if no one has (or given the present state of human capabilities could have) the direct personal experience of bringing about an earthquake by manipulating these plates? Their response to this difficulty is complex, but the central idea is captured in the following passages:

... we would argue that when an agent can bring about one event as a means to bringing about another, this is true in virtue of certain basic intrinsic features of the situation involved, these features being essentially noncausal though not necessarily physical in character. Accordingly, when we are presented with another situation involving a pair of events which resembles the given situation with respect to its intrinsic features, we infer that the pair of events are causally related even though they may not be manipulable. (1993, 197)

Clearly, the agency account, so weakened, allows us to make causal claims about unmanipulable events such as the claim that the 1989 San Francisco earthquake was caused by friction between continental plates. We can make such causal claims because we believe that there is another situation that models the circumstances surrounding the earthquake in the essential respects and does support a means-end relation between an appropriate pair of events. The paradigm example of such a situation would be that created by seismologists in their artificial simulations of the movement of continental plates (1993, 197).

The problem with this strategy parallels the difficulty with von Wright's broadly similar suggestion. What is the nature of the "intrinsic" but (allegedly) "noncausal" features in virtue of which the movements of the continental plates "resemble" the artificial models that the seismologists are able to manipulate? It is well known that small-scale models and simulations of naturally occurring phenomena that superficially resemble or mimic those phenomena may nonetheless fail to capture their causally relevant features because, for example, the models fail to "scale up"—because causal processes that are not represented in the model become quite important at the length scales that characterize the naturally occurring phenomena. Thus, when we ask what it is for a model or simulation that contains manipulable causes to "resemble" phenomena involving unmanipulable causes, the relevant notion of resemblance seems to require that the same *causal* processes are operative in both. Price and Menzies provide no reason to believe that this notion of resemblance can be characterized in noncausal terms. But if the extension of their account to unmanipulable causes requires a notion of resemblance that is already causal in character and which, ex hypothesi cannot be explained in terms of our experience of agency, then their reduction fails.

It might be thought the difficulty under discussion can be avoided by the simple expedient of adhering to a counterfactual formulation of the manipulability theory. Indeed, it is clear that *some* counterfactual formulation is required if the theory is to be even remotely plausible: after all, no one supposes that A can only be a cause of B if A is in fact manipulated. Instead, the intuitive core of the manipulability theory should be formulated as the claim (CF):

(CF) A causes B if and only if B would change if an appropriate manipulation on A *were* to be carried out.

The suggestion under consideration attempts to avoid the difficulties posed by causes that are not manipulable by human beings by contending that for (CF) to be true, it is not required that the manipulation in question be practically possible for human beings to carry out or even that human beings exist. Instead, all that is required is that *if* human beings were to exist and to carry out the requisite manipulation of A (e.g., the continental plates), B (whether or not an earthquake occurs) would change (the possibility of adopting such a counterfactual formulation is sympathetically explored but not fully endorsed by Ernest Sosa and Michael Tooley in the introduction to their 1993 paper).

One fundamental problem with this suggestion is that, independent of whether a counterfactual formulation is adopted, for reasons to be described in section 4, the notion of free action or human manipulation cannot by itself do the work that Menzies and Price wish it to do (that of distinguishing between genuine and spurious causal relationships). But in addition to this, a counterfactual formulation along the lines of (CF) seems completely unilluminating unless accompanied by some sort of account of how we are to understand and assess such counterfactuals and, more specifically, what sort of situation or possibility we are supposed to envision when we imagine that the antecedent of (CF) is true. Consider, for example, a causal claim about the very early universe, during which temperatures are so high that atoms and molecules and presumably anything we would recognize as an agent cannot exist. What counterfactual scenario or possible world are we supposed to envision when we ask, along the lines of (CF), what would happen if human beings were to exist and were able to carry out certain manipulations in this situation? A satisfying version of an agency theory should give us an account of how our experience of agency in ordinary contexts gives us a purchase on how to understand and evaluate such counterfactuals. To their credit, von Wright and Price and Menzies attempt to do this, but in my view, they are unsuccessful.

Causation and Free Action

As we have seen, Menzies and Price assign a central role to "free action" in the elucidation of causation. They do not further explain what they mean by this phrase, preferring instead (as the passage quoted above indicates) to point to a characteristic experience we have as agents. It seems clear, however, that whether (as soft determinists would have it) a free action is understood as an action that is uncoerced or unconstrained or due to voluntary choices of the agent, or whether, as libertarians would have it, a free action is an action that is uncaused or not deterministically caused, the persistence of a correlation between A and B when A is realized as a "free act" is *not* sufficient for A to cause B. Suppose that, in the example described above, the position of the barometer dial X is set by a free act (in either of the above senses) of the experimenter but that that this free act (and hence X) is correlated with Z, the variable measuring atmospheric pressure, perhaps because the experimenter observes the atmospheric pressure and freely chooses to set X in a way that is correlated with Z. (This possibility is compatible with the experimenter's act of setting X being free in either of the above two senses.) In this case, X will remain correlated with Y when produced by a free act, even though X does not cause Y. Suppose, then, that we respond to this difficulty by adding to our characterization of A as realized by a free act the idea that this act must not itself be correlated with any other cause of A (passages in Price (1991) suggest such an additional proviso, although the condition in question seems to have nothing to do with the usual understanding of free action). Even with this proviso, it need not be the case that A causes B if A remains correlated with B when A is produced by an act that is free in this sense, since it still remains possible that the free act that produces A also causes B via a route that does not go through A. As an illustration, consider a case in which an experimenter's administration of a drug to a treatment group (by inducing patients to ingest it) has a placebo effect that enhances recovery, even though the drug itself has no effect on recovery. There is a correlation between ingestion of the drug and recovery that persists under the experimenter's free act of administering the drug even though ingestion of the drug does not cause recovery.

Interventions

Examples like those just described show that if we wish to follow Menzies and Price in defending the claim that if an association between A and B persists when A is given the right sort of "independent causal history" or is "manipulated" in the right way, then A causes B, we need to be much more precise about what we mean by the quoted phrases. There have been a number of attempts to do this in the recent literature on causation. The basic idea that what all of these discussions attempt to capture is a "surgical" change in A which is of such a character that if any change occurs in B, it occurs only as a result of its causal connection, if any, to A and not in any other way. In other words, the change in B, if any, that is produced by the manipulation of A should be produced only via a causal route that goes through A. Manipulations or changes in the value of a variable that have the right sort of surgical features have come to be called *interventions* in the recent literature (e.g., Spirtes, Glymour, and Scheines 1993; Meek and Glymour 1994; Hausman 1998; Pearl [2000] 2009; Woodward 1997, 2000; Woodward and Hitchcock 2001; Cartwright 2003), and I will follow this practice. The characterization of the notion of an intervention is rightly seen by many writers as central to the development of a plausible version of a manipulability theory. One of the most detailed attempts to think systematically about interventions and their significance for understanding causation is due to Pearl (2000) and we turn now to a discussion of his views.

Structural Equations, Directed Graphs, and
Manipulationist Theories of Causation

A great deal of recent work on causation has used systems of equations and directed graphs to represent causal relationships. Judea Pearl (e.g., Pearl [2000] 2009) is an influential example of this approach. His work provides a striking illustration of the heuristic usefulness of a manipulationist framework in specifying what it is to give such systems a causal interpretation.[1] Pearl characterizes the notion of an intervention by reference to a primitive notion of a causal mechanism. A functional causal model is a system of equations $X_i = F(Pa_i, U_i)$ where Pa_i represents the parents or direct causes of X_i that are explicitly included in the model and U_i represents an error term that summarizes the impact of all excluded variables. Each equation represents a distinct causal mechanism that is understood to be "autonomous" in the sense in which that notion is used in econometrics; this means roughly that it is possible to interfere with or disrupt each mechanism (and the corresponding equation) without disrupting any of the others. The simplest sort of intervention in which some variable X_i is set to some particular value x_i amounts, in Pearl's words, to "lifting X_i from the influence of the old functional mechanism $X_i = F_i (Pa_i, U_i)$ and placing it under the influence of a new mechanism that sets the value x_i while keeping all other mechanisms undisturbed" (Pearl [2000] 2009, 70; I have altered the notation slightly). In other words, the intervention disrupts completely the relationship between X_i and its parents so that the value of X_i is determined entirely by the intervention. Furthermore, the intervention is surgical in the sense that no other causal relationships in the system are changed. Formally, this amounts to replacing the equation governing X_i with a new equation $X_i = x_i$, substituting for this new value of X_i in all the equations in which X_i occurs but leaving the other equations themselves unaltered. Pearl's assumption is that the other variables that change in value under this intervention will do so only if they are effects of X_i.

Following Pearl, let us represent the proposition that the value of X has been set by an intervention to some particular value, x_0, by means of a "do" operator $do(X=x_0)$, or more simply, $do(x_0)$. It is important to understand that conditioning on the information that the value of X has been set to x_0 will in general be quite different from conditioning on the information that the value of X has been observed to be x_0 (see Meek and Glymour 1994; Pearl [2000] 2009). For example, in the case in which X and Y are joint effects of the common cause Z, $P(Y/X=x_0) \neq P(Y)$; that is, Y and X are not independent. However, $P(Y/do(X=x_0)) = P(Y)$; that is, Y will be independent of X, if the value of X is set by an intervention. This is because the intervention on

X will break the causal connection from Z to X, so that the probabilistic dependence between Y and X that is produced by Z in the undisturbed system will no longer hold once the intervention occurs. In this way, we may capture Menzies and Price's idea that X causes Y if and only if the correlation between X and Y would persist under the right sort of manipulation of X.

This framework allows for simple definitions of various causal notions. For example, Pearl defines the "causal effect" of X on Y associated with the "realization" of a particular value x of X as:

(C) $P(y/do\ x)$, that is, as the distribution that Y would assume under an intervention that sets the value of X to the value x. It is obvious that this is a version of a counterfactual account of causation.

One of the many attractions of this approach is that it yields a very natural account of what it is to give a causal interpretation to a system of equations of the sort employed in the so-called causal modeling literature. For example, if a linear regression equation $Y = aX + U$ makes a causal claim, it is to be understood as claiming that if an intervention were to occur that sets the value of $X=x_0$ in circumstances $U=u_0$, the value of Y would be $y = ax_0 + u_0$, or alternatively that an intervention that changes X by amount dx will change Y by amount $a\ dx$. As another illustration, consider the system of equations

(1) $Y = aX + U$
(2) $Z = bX + cY + V$
We may rewrite these as follows:
(1) $Y = aX + U$
(3) $Z = dX + W$

where $d = b + ac$ and $W = cU + V$. Since (3) has been obtained by substituting (1) into (2), the system (1)–(2) has exactly the same solutions in X, Y, and Z as the system (1)–(3). Since X, Y and Z are the only measured variables, (1)–(2) and (1)–(3) are "observationally equivalent" in the sense that they imply or represent exactly the same facts about the patterns of correlations that obtain among the measured variables. Nonetheless, the two systems correspond to different causal structures. (1)–(2) says that X is a direct cause of Y and that X and Y are direct causes of Z. By contrast, (1)–(3) says that X is a direct cause of Y and that X is a direct cause of Z but says nothing about a causal relation between Y and Z. We can cash this difference out within the interventionist/manipulationist framework described above—(2) claims that an intervention on Y will change Z while (3) denies this (recall that an intervention on Y with respect to Z must not be correlated with any other cause of Z such as X, and will break any causal connection between X and Y). Thus while the two systems of equations agree about the correlations so far observed, they disagree about what would happen under an intervention on Y. According to an interventionist/manipulationist account of causation, it is the system that gets such counterfactuals right that correctly represents the causal facts.

One possible limitation of Pearl's characterization of an intervention concerns the scope of the requirement that an intervention on X_i leave intact *all* other mechanisms besides the mechanism that previously determined the value of X_i. If, as Pearl apparently intends, we understand this to include the requirement that an intervention on X_i must leave intact the causal mechanism (if any) that connects X_i to its possible effects Y, then an obvious concern about circularity arises, at least if we want to use the notion of an intervention to characterize what it is for X_i to cause Y. A closely related problem is that given the way Pearl characterizes the notion of an intervention, his definition (C) of the causal effect of X on Y, seems to give us not the causal contribution made by $X = x$ alone to Y but rather the combined impact on Y of this contribution and whatever contribution is made to the value of Y by other causes of Y besides X. For example, in the case of the regression equation $Y = aX + U$, the causal effect in Pearl's sense of $X=x$ *on* Y is apparently $P(Y) = ax + U$, rather than, as one might expect, just ax. In part for these reasons, Woodward (2003) and Woodward and Hitchcock (2003) explore a different way of characterizing the notion of an intervention that does not make reference to the relationship between the variable intervened on and its effects. For Woodward and Hitchcock, in contrast to Pearl, an intervention I on a variable X is always defined with respect to a second variable Y (the intent being to use the notion of an intervention on X with respect to Y to characterize what it is for X to cause Y). Such an intervention I must meet the following requirements (M1)–(M4):

(M1) I must be the only cause of X (i.e., as with Pearl, the intervention must completely disrupt the causal relationship between X and its previous causes so that the value of X is set entirely by I),

(M2) I must not directly cause Y via a route that does not go through X as in the placebo example,

(M3) I should not itself be caused by any cause that affects Y via a route that does not go through X, and

(M4) I leaves the values taken by any causes of Y unchanged except those that are on the directed path from I to X to Y (should this exist).

Within this framework, the most natural way of defining the notion of causal effect is in terms of the *difference* made to the value of Y by a change or difference in the value of X. Focusing on differences in this way allows us to isolate the contribution made to Y by X alone from the contribution made to Y by its other causes. Moreover, since in the nonlinear case, the change in the value of Y caused by a given change in the value of X will depend on the values of the other causes of Y, it seems to follow that the notion of causal effect must be relativized to a background context B_i that incorporates information about these other values. In deterministic contexts, we might thus define the causal effect on Y of a change in the value of X from X=x to X=x' in circumstances B_i as:

(CD) $Y_{do\, x,\, Bi} - Y_{do\, x',\, Bi'}$ that is, as the difference between the value that Y would take under an intervention that sets X=x in circumstances B_i and the value that Y would take under an intervention that sets X=x' in $B_{i'}$ where the notion of an intervention is now understood in terms of (M1)–(M4) rather than in the way recommended by Pearl. In nondeterministic contexts, the characterization of causal effect is less straightforward, but one natural proposal is to define this notion in terms of expectations: if we let $E\, P\, do\, x$, and $B_i(Y)$ be the expectation of Y with respect to the probability distribution P if X is set to X=x by means of an intervention, then the causal effect on Y of a change in X from X=x" to X=x might be defined as: $E\, P\, do\, x$, $B_i(Y) - E\, P\, do\, x'$, $B_i(Y)$. In the deterministic case, X will then be a cause of Y in B_i if and only if the causal effect of X on Y in B_i is nonzero for some pair of values of X—that is, if and only if there are distinct values of X, x and x' such that the value of Y under an intervention that sets X=x in B_i is different from the value of Y under an intervention that sets X=x'. In probabilistic contexts, X will be a cause of Y if the expectation of Y is different for two different values of X, when these are set by interventions.

We will not attempt to adjudicate here among these and various other proposals concerning the best way to characterize the notions of intervention and causal effect. Instead, we want to comment on the general strategy they embody and to compare it with the approach to causation associated with theorists like Menzies and Price. Note first that the notion of an intervention, when understood along either of the lines described above, is an unambiguously causal notion in the sense that causal notions are required for its characterization—thus the proposals variously speak of an intervention on X as breaking the causal connection between X and its causes while leaving other causal mechanisms intact, or as not affecting Y via a causal route that does not go through X. This has the immediate consequence that one cannot use the notion of an intervention to provide a reduction of causal claims to noncausal claims. Moreover, to the extent that reliance on some notion like that of an intervention is unavoidable in any satisfactory version of a manipulability theory (as I believe that it is), any such theory must be nonreductionist. Indeed, we can now see that critics who have charged manipulability theories with circularity have in one important sense understated their case: manipulability theories turn out to be "circular" not just in the obvious sense that for an action or event I to constitute an intervention on a variable X, there must be a causal relationship between I and X, but in the sense that I must meet a number of other causal conditions as well.

Is Circularity a Problem?

Suppose that we agree that any plausible version of a manipulability theory must make use of the notion of an intervention and that this must be characterized in causal terms. Does this sort of "circularity" make any such theory trivial and unilluminating? It seems to me that it does not, for at least two reasons. First, it may be, as writers like Woodward contend (2003), that in characterizing what it is for a process I to qualify as an intervention on X for the purposes of characterizing what it is for X to cause Y, we need not make use of information about the causal relationship, if any, between X and Y. Instead, it may be that we need only to make use of *other* sorts of causal information (e.g., about the causal relationship between I and Y or about whether I is caused by causes that cause Y without causing X), as in (M1)–(M4) above. To the extent that this is so, we may use one set of claims about causal relationships (e.g., that X has been changed by a process that meets the conditions for an intervention) together with correlational information (that X and Y remain correlated under this change) to characterize what it is for a different relationship to be causal (the relationship between X and Y). This does not yield a reduction of causal talk to noncausal talk, but it is also not viciously circular in the sense that it presupposes that we already have causal information about the very relationship that we are trying to characterize. One reason for thinking that there must be *some* way of characterizing the notion of an intervention along the lines just described is that we do sometimes learn about causal relationships by performing experiments—and it is not easy to see how this is possible if in order to characterize the notion of an intervention on X we had to make reference to the causal relationship between X and its effects.

A related point is that even if manipulability accounts of causation are nonreductive, they can *conflict* with other accounts of causation, leading to different causal judgments in particular cases. As an illustration, consider a simple version of manipulability account along the lines of (CD), according to which a sufficient condition for X to cause (have a causal effect on Y) is that some change in the value of X produced by an intervention is associated with a change in the value of Y. Such an account implies that omissions can be causes (e.g., the failure of a gardener to water a plant causes the plant's death) since a change under an intervention in whether the gardener waters is associated with a change in the value of the variable measuring whether the plant dies. For a similar reason, relationships involving "double prevention" (Hall 2000) or "causation by disconnection" (Schaffer 2000) count as genuine causal relationships on interventionist accounts. Consider, by contrast, the verdicts about these cases reached by a simple version of a *causal process* theory (in the sense of Salmon 1984 and Dowe 2000) according to which a necessary condition for a particular instantiation x of a value X to cause a particular instantiation y of a value Y is that there be a spatiotemporally continuous process connecting x to y involving the transfer of energy, momentum, or perhaps some other conserved quantity. According to such a theory, "causation" by omission or by double prevention does not qualify as genuine causation. Similarly, if an "action at a distance" version of Newtonian gravitational theory had turned out to be correct, this would be a theory that described genuine causal relationships according to interventionist accounts of causation, but not according to causal process accounts. Whether one regards the verdicts about these cases reached by causal process accounts or by interventionist accounts as more defensible, the very fact that the accounts lead to inconsistent judgments shows that interventionist approaches are not trivial or vacuous, despite their "circular," nonreductive character.

The Plurality of Causal Concepts

A second respect in which reliance on the notion of an intervention need not be thought of as introducing a vicious circularity is this: so far, we have been following von Wright and Menzies and Price in assuming that there is just one causal notion or locution (A causes B, where A and B are types of events) that we are trying to analyze. But in fact there are many such notions. For example, among causal notions belonging to the family of so-called type causal notions (i.e., causal claims that relate types of events or variables) there is a distinction to be drawn between what we might call claims about total or net causes and claims about direct causes. Even if the notion of an intervention presupposes some causal notion such as some notion of type causation, it may be that we can use it to characterize other causal notions.

As an illustration, consider the causal structure represented by the following equations and associated directed graph:

$Y = aX + cZ$
$Z = bX$

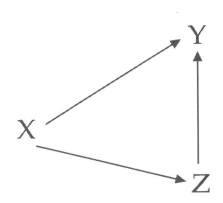

In this structure, there are two different causal routes from X to Y—a direct causal relationship and an indirect relationship with Z as an intermediate variable. If $a = -bc$, there is cancellation along these two routes. This means that no intervention on X will change the value of Y. In one natural sense, this seems to mean that X does not cause Y, just as (C) (section 6) suggests. In another natural sense, however, X does seem to be a cause—indeed a direct cause—of Y. We can resolve this apparent inconsistency by distinguishing between two kinds of causal claims[2]— the claim X is a total or net cause of Y, where this is captured by (C) or (CD), and the claim that X is a direct cause of Y, where this is understood along the following lines: X is a direct cause of Y if and only if under some intervention that changes the value of X, the value of Y changes when all other variables in the system of interest besides X and Y, including those that are on some causal route from X to Y, are held fixed at some value, also by interventions (for related, but different, characterizations of direct causation along these lines, see Pearl ([2000] 2009) and Woodward (2003)). Fixing the other values of other variables means that each of these values are determined by separate processes, each meeting the conditions for an intervention, that are appropriately independent of each other and of the intervention that changes the value of X. Thus, the effect of intervening to fix the values of these variables is that each variable intervened on is disconnected from its causes, including X. In the example under discussion, X qualifies as a direct cause of Y because if we were to fix the value of Z

in a way that disconnects it from the value of X, and then intervene to change the value of X, the value of Y would change. This idea can then be generalized to provide a characterization of "contributing" causation along a causal route (i.e., to capture the sense in which X is an indirect cause of Y along the route that goes through Z) (Woodward 2003).

So far our focus has been on type causal claims of various kinds. There are also a number of proposals in the literature that provide interventionist treatments of token or actual cause claims (these have to do with the event of X's taking on a particular value being an actual cause of Y's taking on a particular value), including those that involve various forms of preemption and over-determination (e.g., Halpern and Pearl 2001; C. Hitchcock 2001a; Woodward 2003; C. Hitchcock 2007). Considerations of space preclude detailed description, but one strategy that has been explored is to appeal to what will happen to the effect under *combinations* of interventions that both affect the cause and that fix certain other variables to specific values. As an illustration, consider a standard case of causal preemption: Gunman one shoots (s_1) victim, causing his death d, while gunman two does not shoot but would have shot (s_2) also causing d, if s_1 had not occurred. If we fix (via an intervention) the behavior of the gunman two at its actual value (he does not shoot), then an independent intervention that alters whether gunman one shoots will alter whether victim dies, thus identifying s_1 as the actual cause of d, despite the absence of counterfactual dependence (of the usual sort) between d and s_1. Accounts along these lines are able to deal with a number (although admittedly not all[3]) of the standard counterexamples to other counterfactual treatments of token causation.

It is worth adding that although this appeal to combinations of interventions may seem artificial, it maps on to standard experimental procedures in an intuitive way. Consider a case of genetic redundancy— gene complex G_1 is involved in causing phenotypic trait P but if G_1 is inactivated another gene complex G_2 (which is inactive when G_1 is active) will become active and will cause P. The geneticist may test for this possibility by first intervening on G_2 so that it is fixed at the value "inactive," then intervening to vary G_1 and observing whether there is a corresponding change in P. Second, the investigator may intervene to render G_1 inactive and then, independently of this, intervene to change G_2 and observe whether there is a change in P. As this example illustrates, we may think of different complex causal structures in which there are multiple pathways, redundancy, cancellation, and so on, as encoding different sets of claims about what will happen under various possible combinations of interventions.

Thus even if a "manipulationist" or "interventionist" framework does not yield a reduction of causal talk to noncausal talk, it provides a natural way of marking the distinctions among a number of different causal notions and exhibiting their interrelations. More generally, even if a manipulationist account of causation does not yield a reduction but instead simply connects "causation" (or better, various more specific causal concepts) with other concepts within the same circle, we still face many nontrivial choices about how the concepts on this circle are to be elucidated and connected with one another. For example, it is far from obvious how to characterize the notion of an intervention so as to avoid the various counterexamples to standard statements of the manipulability theory, such as the theory of Menzies and Price. It is in part because the notion of manipulation/intervention has an interesting and complex fine structure—a structure that is left largely unexplored in traditional manipulability theories—that working out the connection between causation and manipulation turns out to be interesting and nontrivial.

Interventions That Do Not Involve Human Action

We noted above that a free action need not meet the conditions for an intervention on any of the conceptions of intervention described in section 6. It is also true that a process or event can qualify as an intervention even if it does not involve human action or intention at any point. This should be apparent from the way the notion of an intervention has been characterized, for this is entirely in terms of causal and correlational concepts and makes no reference to human beings or their activities. In other words, a purely "natural" process involving no animate beings at all can qualify as an intervention as long as it has the right sort of causal history—indeed, this sort of possibility is often described by scientists as a natural experiment. Moreover, even when manipulations are carried out by human beings, it is the causal features of those manipulations and not the fact that they are carried out by human beings, or are free, or are attended by a special experience of agency, that matters for recognizing and characterizing causal relationships. Thus, by giving up any attempt at reduction and characterizing the notion of an intervention in causal terms, an "interventionist" approach of the sort described under sections 5 and 6 avoids the second classical problem besetting manipulability theories—that of anthropocentrism and commitment to a privileged status for human action. For example, under this approach X will qualify as a (total) cause of Y as long as it is true that for some value of X that if X were to be changed to that value by a process having the right sort of causal characteristics, the value of Y would change. Obviously, this claim can be true even if human beings lack the power to manipulate X or even in a world in which human beings do not or could not exist. There is nothing in the interventionist version of a manipulability theory that commits us to the view that all causal claims are in some way dependent for their truth on the existence of human beings or involve a "projection" onto the world of our experience of agency.

Interventions and Counterfactuals

We noted above that interventionist versions of manipulability theories are counterfactual theories. What is the relationship between such theories and more familiar versions of counterfactual theories such as the theory of David Lewis? Lewis's theory is an account of what it is for one individual token event to cause another, while (C) is formulated in terms of variables or types of events, but abstracting away from this and certain other differences, there are a number of striking similarities between the two approaches. As readers of Lewis will be aware, any counterfactual theory must explain what we should envision as changed and what should be held fixed when we evaluate a counterfactual, the antecedent of which is not true of the actual world—within Lewis's framework, this is the issue of which worlds in which the antecedent of the counterfactual holds are "closest" or "most similar" to the actual world. Lewis's answer to this question invokes a "similarity" ordering that ranks the importance of various respects of resemblance between worlds in assessing overall similarity (Lewis 1979). For example, avoiding diverse, widespread violations of law is said to be the most important consideration, preserving perfect match of particular fact over the largest possible spatiotemporal region is next in importance and more important than avoiding small localized violations of law, and so on. As is well known, the effect of this similarity ordering is, at least in most situations, to rule out so-called backtracking counterfactuals (e.g., the sort of counterfactual that is involved in reasoning that if the effect of some cause had not occurred, then the cause would not have occurred). When the antecedent of a counterfactual is not true of the actual world, Lewis's similarity metric leads us (at least in deterministic contexts) to think of that antecedent as made true by a "small" miracle.

The notion of an intervention plays a very similar role within manipulability theories that causation does in Lewis's similarity ordering. Like Lewis's ordering, the characterization of an intervention tells us what should be envisioned as changed and what should be held fixed when we evaluate a counterfactual like, "If X were to be changed by an intervention to such and such a value, the value of Y would change." For example, based on Pearl's understanding of an intervention, in evaluating this counterfactual, we are to consider a situation in which the previously existing causal relationship between X and its causes is disrupted, but all other causal relationships in the system of interest are left unchanged. A moment's thought will also show that, as in Lewis's account, both Pearl's and Woodward's characterizations of interventions rule out backtracking counterfactuals—for example, in evaluating a counterfactual of the form "if an intervention were to occur that changes E (where E is an effect of C), then C would change," Pearl holds that we should consider a situation in which the relationship between E and its causes (in this case, C) is disrupted but all other causal relationships are left unchanged, so that C still occurs, and the above counterfactual is false, as it should be. Moreover, there is a clear similarity between Lewis's idea that the appropriate counterfactuals for analyzing causation are counterfactuals the antecedents of which are made true by miracles, and the idea of an intervention as an exogenous change that disrupts the mechanism that was previously responsible for the cause event C. Indeed, one might think of an interventionist treatment of causation as explaining why Lewis's account with its somewhat counterintuitive similarity ordering works as well as it does—Lewis's account works because his similarity ordering picks out roughly those relationships that are stable under interventions and hence exploitable for purposes of manipulation and control and, as a manipulability theory claims, it is just these relationships that are causal. This is not to say, however, that the two approaches always yield identical assessments of particular causal and counterfactual claims—Woodward (2003) describes cases in which the two approaches diverge and in which the interventionist approach seems more satisfactory[4].

Possible and Impossible Interventions

In the version of a manipulability theory considered above under section 6, causal claims are elucidated in terms of counterfactuals about what would happen under interventions. As we have seen, the notion of an intervention should be understood without reference to human action, and this permits formulation of a manipulability theory that applies to causal claims in situations in which manipulation by human beings is not a practical possibility. Moreover, the counterfactual formulation allows us to make sense of causal claims in contexts in which interventions do not in fact occur and arguably even in cases in which they are causally impossible, as long as we have some principled basis to answer questions about what *would* happen to the value of some variable *if* an intervention were to occur on another variable. Consider, for example, the (presumably true) causal claim (G):

(G) The gravitational attraction of the moon causes the motion of the tides.

Human beings cannot at present alter the attractive force exerted by the moon on the tides (e.g., by altering its orbit). More interestingly, it may well be that there is no physically possible process that will meet the conditions for an intervention on the moon's position with respect to the tides—all possible processes that would alter the gravitational force exerted by the moon may be insufficiently "surgical." For example, it may very well be that any possible process that alters the position of the moon by altering the position of some other massive object will have an independent impact on the tides in violation of condition (M2) for an intervention. It is nonetheless arguable that we have a principled basis in Newtonian mechanics and gravitational theory themselves for answering questions about what would happen if such a surgical intervention were to occur and that this is enough to vindicate the causal claim (G).

Although this strategy of appealing to counterfactuals about what would happen under interventions that may not be causally possible helps to address some concerns that interventionist accounts are too narrow in scope in the sense of failing to capture some causal claims like (G) that seem scientifically well founded, it seems clear that as we make the relevant notion of "possible intervention" more and more permissive so that it includes various sorts of contra-nomic possibilities, we will reach a point at which this notion and the counterfactuals in which it figures become so unclear that we can no longer use them to illuminate or provide any independent purchase on causal claims. It is an interesting and unresolved question whether the point at which the associated causal claims no longer strike us as clear or useful, which is what one would expect if interventionism is a complete account of causation.

Scope of Interventionist Accounts

This issue arises in a particularly forceful way when we attempt to apply such accounts to fundamental physical theories understood as applying to the whole universe. Consider the following claim:

(12.1) The state S_t of the entire universe at time t causes the state S_{t+d} of the entire universe at time $t+d$, where S_t and S_{t+d} are specifications in terms of some fundamental physical theory.

On an interventionist construal, (12.1) is unpacked as a claim to the effect that under some possible intervention that changes S_t, there would be an associated change in S_{t+d}. The obvious worry is that it is unclear what would be involved in such an intervention and unclear how to assess what would happen if it were to occur, given the stipulation that S_t is a specification of the entire state of the universe. How, for example, might such an intervention be realized, given that there is nothing left over in addition to S_t to realize it with?

Commenting on an example like this, Pearl writes:

> If you wish to include the whole universe in the model, causality disappears because interventions disappear—the manipulator and the manipulated lose their distinction. ([2000] 2009, 350)

Whether or not Pearl is right about this, it seems uncontroversial that it is far from straightforward how to interpret the interventionist counterfactual associated with (12.1). The interventionist account seems to apply most naturally and straightforwardly to what Pearl calls "small worlds"—cases in which the system of causal relationships in which we are interested is located in a larger environment that serves as a potential source of outside or "exogenous" interventions. The systems of causal relationships that figure in common sense causal reasoning and in the biological, psychological, and social sciences all have this character but fundamental physical theories do not, at least when their domain is taken to be the entire universe.

There are several possible reactions to these observations. One is that causal claims in fundamental physics like (12.1) are literally true and that it is an important limitation of interventionist theories that they have difficulty elucidating such claims. A second, diametrically opposed reaction, which we take to be Pearl's, is that causal concepts do not apply, at least in any straightforward way, to some or many fundamental physics contexts and that it is a virtue of the interventionist account that it helps us to understand why this is so. This second suggestion may seem deeply counterintuitive to philosophers who believe that fundamental physical laws should be understood as making causal claims that serve to "ground" true causal claims made by common sense and the special sciences. In fact, however, the view that fundamental physics is not a hospitable context for causation and that attempts to interpret fundamental physical theories in causal terms are unmotivated, misguided, and likely to breed confusion is probably the dominant, although by no means universal, view among contemporary philosophers of physics[5]. According to some writers (H. Hitchcock 2007; Woodward 2007), we should take seriously the possibility that causal reasoning and understanding apply most naturally to small world systems of medium-sized physical objects of the sort studied in the various special sciences and look for an account of causation, like the interventionist account, that explains this fact. The question of the scope of interventionist theories and their implications for causal claims in fundamental physics is thus an important and at present unresolved issue.[6]

(Alleged) Causes That Are Unmanipulable for Logical, Conceptual, or Metaphysical Reasons

Several statisticians (e.g., Holland 1986; Rubin 1986) who advocate manipulationist or counterfactual accounts of causation have held that causal claims involving causes that are unmanipulable in principle are defective or lack a clear meaning—they think of this conclusion as following directly from a manipulationist approach to causation. What is meant by an unmanipulable cause is not made very clear, but the examples discussed typically involve alleged causes (e.g., race, or membership in a particular species, or perhaps gender) for which we lack any clear conception of what would be involved in manipulating them or any basis for assessing what would happen under such a manipulation. Such cases contrast with the case involving (G) above, where the notion of manipulating the moon's orbit seems perfectly clear and well defined, and the problem is simply that the world happens to be arranged in such a way that an intervention that produces such a change is not physically possible.

A sympathetic reconstruction of the position under discussion might go as follows. On a manipulationist account of causation, causes (whether we think of them as events, types of events, properties, facts, or what have you) must be representable by means of *variables*—where this means, at a minimum, that it must be possible for the cause to change or to assume different values. This is required if we are to have a well-defined notion of manipulating a cause and well-defined answers to counterfactual queries about what would happen if the cause were to be manipulated in some way—matters that are central to what causal claims mean on any version of a manipulability theory worthy of the name. Philosophers tend to think of causes as properties or events, but in many cases, it is straightforward to move back and forth between such talk and a representation in terms of variables, as we have been doing throughout this entry. For example, rather than saying that the impact of the baseball caused the window to shatter or that impacts of baseballs cause window shatterings, we may introduce two indicator variables—*I*, which takes the values 0 and 1 for {no impact, impact}, and *S*, which takes the values 0 and 1 for {no shattering, shattering}, and use these variables to express the idea that whether or not the window shatters is counterfactually dependent on interventions that determine whether the impact occurs. Both *I* and *S* describe causes that are straightforwardly manipulable. However, for some putative causes, there may be no well-defined notion of change or variation in value, and if so, a manipulability theory will not count these as genuine causes. For example, if it is metaphysically necessary that everything that exists is a physical object or if we lack any coherent conception of what it is for something to exist but to be nonphysical, then there will be no well-defined notion of intervening to change whether something is a physical object. While there are true (and even lawful) generalizations about all physical objects, on a manipulability theory these will not describe causal relationships. Thus, although to the best of our knowledge, it is a law of nature that (L) no physical object can be accelerated from a velocity less than that of light to a velocity greater than light, (L) is not, according to a manipulability theory, a *causal* generalization.

Moreover, even with respect to variables that can take more than one value, the notion of an intervention or manipulation will not be well defined if there is no well-defined notion of *changing* the values of that variable. Suppose that we introduce a variable "animal" that takes the values {lizard, kitten, raven}. By construction, this variable has more than one value, but if, as seems plausible, we have no coherent idea of what it is to change a raven into a lizard or kitten, there will be no well-defined notion of an intervention for this variable, and being an animal (or being a raven) will not be the sort of thing that can count as a bona-fide cause on a manipulability theory. The notion of changing the value of a variable seems to involve the idea of an alteration from one value of the variable to another in circumstances in which the very same system or entity can possess both values and this notion seems inapplicable to the case under discussion.

Some readers will take it to be intuitively obvious that being a raven can be a cause (e.g., of some particular organism's being black). Many standard theories of causation also endorse this conclusion; for example, if we are willing to assume it is a law that all ravens are black, then nomological theories of causation will support the claim (R):

134

(R) Ravenness causes blackness.

Similarly, ravenness raises the probability of blackness and hence (R) qualifies as causal on probabilistic theories of causation, and depending on how the relevant similarity ordering is understood, (R) may also qualify as causal on a Lewis style counterfactual theory. If causal claims like (R) are true, it is an important inadequacy in manipulability theories that they seem unable to capture such claims. By contrast, others will think that claims like (R) are, if not false, at least unclear and unperspicuous, and that it is a point in favor of manipulability theories that they explain why this is the case. Those who take this second view will think that claims like (R) should be replaced by claims that involve causes that are straightforwardly manipulable. For example, (R) might be replaced by a claim that identified the genetic factors and biochemical pathways that are responsible for raven pigmentation—factors and pathways for which there is a well-defined notion of manipulation and which are such that if they were appropriately manipulated, would cause changes in pigmentation. Manipulability theorists like Rubin and Holland will think that such a replacement would be clearer and more perspicuous than the original claim (R). In any case, claims involving causes that are unmanipulable in the sense that we seem to lack any clear conception of what would be involved in manipulating them are one important sort of case in which a manipulability approach will diverge from many other standard theories of causation.

Consider an additional illustration of this general theme. Holland (1986) appeals to a manipulability theory of causation to argue that the following claim is fundamentally unclear:

(F) Being female causes one to be discriminated against in hiring and (or) salary.

In contrast to the previous cases, the problem here is not so much that under all interpretations of the putative cause ("being female") we lack any clear idea of what it would be like to manipulate it but rather that there are several rather different things that might be meant by manipulation of "being female" (which from the perspective of a manipulability theory is to say that there several quite different variables we might have in mind when we talk about being female as a cause), and the consequences for discrimination of manipulating each of these may be quite different. For example, (F) might be interpreted as claiming that a literal manipulation of gender, as in a sex change operation, that leaves an applicant's qualifications otherwise unchanged, will change expected salary or probability of hiring. Alternatively, and more plausibly, (F) might be interpreted as claiming that manipulation of a potential employer's *beliefs* about an applicant's gender will change salary and hiring probability, in which case (F) would be more perspicuously expressed as the claim that employer beliefs about gender cause discrimination. Still another possible interpretation—in fact what Holland claims one *ought* to mean by (F)—is that differentials in salary and hiring between men and women would disappear (or at least be reduced substantially) under a regime in which various sorts of biased practices were effectively eliminated, presumably as the result of changes in law and custom. While we see no reason to follow Holland in thinking that this is the only legitimate interpretation of (F), it is plainly a legitimate interpretation. Moreover, Holland is also correct to think that this last hypothetical experiment, which involves manipulating the legal and cultural framework in which discrimination takes place, is a quite different experiment from an experiment involving manipulating gender itself or employee beliefs about gender and that each of these experiments is likely to lead to different outcomes. From the perspective of a manipulability theory, these different experiments thus correspond to different causal claims. As this example illustrates, part of the heuristic usefulness of a manipulability theory is that it encourages us to clarify or disambiguate causal claims by explicitly distinguishing among different possible claims regarding the outcomes of hypothetical experiments that might be associated with them. That we can clarify the meaning of a causal claim in this way is just what we would expect if a manipulability account of causation were correct.

More Recent Criticisms of Interventionist Accounts

A number of other criticisms besides the classic charges of anthropomorphism and circularity have been advanced against interventionist accounts. One complaint is that interventionist accounts (at least as we have formulated them) appeal to counterfactuals and that counterfactuals cannot be (as it is often put) "barely true": if a counterfactual is true, this must be so in virtue of some "truth maker" which is not itself modal or counterfactual. Standard candidates for such truth makers are fundamental laws of nature or perhaps fundamental physical/chemical processes or mechanisms. Often the further suggestion is made that we can then explain the notion of causation in terms of such truth makers rather than along interventionist lines—for example, the notion of causation (as well as the truth conditions for counterfactuals) might be explained in terms of laws (Hiddleston 2005). Thus appealing to

SUBSTANTIVE THEORY AND CONSTRUCTIVE MEASURES

interventionist counterfactuals is not necessary, once we take account of the truth conditions of such counterfactuals.

These claims raise a number of issues that can be explored only briefly. First, let us distinguish between providing an ordinary scientific explanation for why some counterfactual claim is true and providing truth conditions (or identifying a truth maker) in the sense described above, where these truth conditions are specified in nonmodal, noncounterfactual terms. The expectation that (i) *whenever some macro-level interventionist counterfactual is true, there will be some more fundamental scientific explanation of why it is true* seems plausible and well grounded in scientific practice. By contrast, the expectation that (ii) *for every true counterfactual there must be a truth maker that can be characterized in noncounterfactual terms* is a metaphysical doctrine that requires some independent argument; it does not follow just from (i). Suppose that it is true that (14.1) if subjects with disease *D* were to be assigned treatment via an intervention with drug *G*, they would be more likely to recover. Then it is very plausible that there will be some explanation, which may or may not be known at present, that explains why (14.1) is true in terms of more fundamental biochemical mechanisms or physical/chemical laws and various initial and boundary conditions. What is less obviously correct is the further idea that we can elucidate these underlying mechanisms/laws without appealing to counterfactuals. It is this further idea that is appealed to when it is claimed that it must be possible to describe a truth maker for a counterfactual like (14.1) that does not itself appeal to counterfactual or modal claims. The correctness of this idea is not guaranteed merely by the existence of an explanation in the ordinary sense for why (14.1) is true; instead it seems to depend on whether a reductivist account of laws, mechanisms, and so on in terms of nonmodal primitives can be given—a matter on which the jury is still out.[2]

A different line of criticism has been advanced against interventionist accounts in several recent papers by Nancy Cartwright (e.g., 2001, 2002). According to Cartwright, such accounts are "operationalist." Classical operationalism is often criticized as singling out just one possible procedure for testing some claim of interest and contending that the claim only makes sense or only has a truth value when that procedure can actually be carried out. Similarly, Cartwright complains that the interventionist account "overlooks the possibility of devising other methods for measuring" causal relationships and also suggests that the account leads us to "withhold the concept [of cause] from situations that seem the same in all other aspects relevant to its application just because our test cannot be applied in those situations" (2002, 422).

If interventionism is formulated as above, this criticism seems misplaced. The interventionist account does not hold that causal concepts apply or make sense only when the appropriate interventions can actually be carried out. Nor does it deny that there are other ways of testing causal claims besides carrying out interventions. Instead, interventionism holds that causal claims apply or have truth values whenever the appropriate counterfactuals concerning what would happen if interventions were to be performed have truth values. As explained above, interventionists think that sometimes such counterfactuals are true even if the interventions in question cannot actually be performed. Similarly, interventionists can readily agree that causal claims may be tested and confirmed by, for example, purely observational data not involving interventions or manipulations—their view, though, is that what is confirmed in this way is a claim about what would happen if certain interventions were to be performed.

In a related criticism, Cartwright contends that the interventionist account is "monolithic": it takes just one of the criteria commonly thought to be relevant to whether a relationship is causal—whether it is potentially exploitable for purposes of manipulation—and gives it a privileged or preeminent place, allowing it to trump other criteria (like spatiotemporal contiguity or transmission of energy momentum) when it comes into conflict with them. By contrast, Cartwright favors a "pluralistic" account, according to which a variety of diverse criteria are relevant to whether a relationship is causal, and determining which of these are most appropriate or important will depend on the causal claim at issue.

The interventionist account is indeed mono-criterial. Whether this feature is objectionable depends on whether there are realistic cases in which (i) intervention-based criteria and criteria based on other considerations come into conflict *and* (ii) it is clear that the causal judgments supported by these other criteria are more defensible than those supported by interventionist criteria. Cartwright does not present any uncontroversial cases of this kind. We have seen that interventionist accounts that take, for example, spatiotemporal continuity to be crucial for causation do yield conflicting judgments in some realistic cases (e.g., those involving double prevention), but it is far from clear that the interventionist account is mistaken in the judgments that it recommends about such cases.

Box 16: Concatenating Sticks and Measurement

Although Ben Wright sometimes used the analogy of sticks when discussing measures, in *Best Test Design* (Wright and Stone 1979), we settled on the arrow as a useful analogy. Our later work *Measurement Essentials* (Wright and Stone 1996) included a figure encompassed by sticks, and the sticks have since come to serve as a *primitive measure* analogy used to illustrate constructing variables. This *urgrund* [very basis] of sticks becomes a foundation for making measures. The sequence is as follows:

1. Comparison is key (i.e., equal, greater, less).
2. Compare a chosen stick (agent) to an object. The stick is equal to, greater, or less.
3. If greater than, choose a longer stick and compare again; equal to, greater, or less.
4. If less than, choose a shorter stick and compare; equal to, greater, or less.
5. Select as many sticks (agents) as required, longer or shorter, to "box in" the object.
6. When the sticks chosen for use are selected and ordered with an equal difference between each of them, you have created a unit.
7. The more sticks chosen, the finer the difference (unit) by which to box in the object.
8. Assemble the sticks in their graduated order. Comparison and order are shown to be essential to this process.
9. If one could collapse all the individual sticks into a single one, instead of an assemblage of sticks there would be a "ruler" constructed from the sticks.
10. Assigning numbers to this orderly sequence of sticks produces a "stick ruler" similar to a foot ruler demarcated by inches, or a yardstick by inches/feet, rod by feet, mile by feet, rods, and so forth.

The analogy of sticks provokes the construction of spelling items, math problems, indicants of depression, fear, anxiety, shame, and so forth. This approach is further illustrated and confirmed by observing the decreasing lengths of marimba key length for unit scale differences in well-tempered pitch, for the varying string lengths observed on a harp or concert grand piano, and the decreasing lengths of organ pipes. It is further manifested by experiments dating at least as far back as the Greeks, who used the monochord as the basis for the study of "music" (more correctly, the fundamentals of today's "acoustics") comprising one science in the medieval quadrivium [arithmetic, geometry, music, and cosmology].

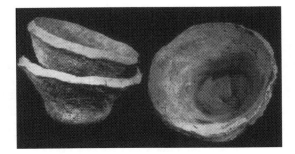

According to archeological findings, unique aspects for developing measures were employed much earlier. One example from the Early High period of the ancient Middle East (c. 3200–2800 BCE) is documented by Nissen (1988), who identified a "unit bowl" used for dispensing the daily food ration across the entire Babylonian empire. The proliferation of these unit bowls throughout the empire testifies to a standard economic unit identifiable not just by its unit volume but by its unique construction, which designates its singular purpose and keeps it distinct from all other pottery. As shown above, the pictograph of this bowl together with a head indicates "to eat."

An orderly arrangement of sticks can be associated with the axioms of quantity suggested by Hölder (1901) as given by Nagel (1931):

Nagel's axioms	Stick order as numbered above
1. Either a > b, or a < b, or a = b.	1, 2, 3, 4
2. If a > b, and b > c, then a > c.	5
3. For every a there is an a' such that a = a'.	2, 3, 4
4. If a > b, and b = b', then a > b'.	2, 3, 4
5. If a = b, then b = a.	1, 2, 3, 4
6. For every a there is a b such that a > b (within limits).	2, 3, 4
7. For every a and b there is a c such that c = a + b.	8, 9
8. a + b > a'.	8, 9
9. a + b = a' + b'.	2, 3, 4

10. a + b = b + a.	1
11. (a + b) + c = a + (b + c).	8, 9
12. If a < b, there is a number n such that na > b (also within limits).	8, 9

These axioms can be somewhat aligned to the sequence of sticks as indicated in the column to the right of the axioms. The association need not correspond perfectly to substantiate the value of the "sticks" analogy in demonstrating how useful a primitive form of determining measures can be in producing a model for measuring. The alignment is not nearly as important as the fact that the sticks and axioms are both orderly constructions illustrating their importance in systematic fabrication. Order by comparison is essential to any construction, or as Luce and Narens write, "All measurement rests upon having a qualitative ordering of the set of objects" (1981, 215).

We should not slight or dismiss the role of simple strategies for conveying sophisticated processes. Once considered the domain of analysis, elementary school students are now exposed to previously advanced concepts such as the commutative, associative, distributive, and transitive properties. Leopold Kronecker is said to have attributed creation of the integers to God, and creation of all other mathematics to humankind. We are, and have long been, a fabricating people.

Michell provides some similar conditions that characterize length as a continuous quantitative variable:

1. For every pair of lengths, a and b, one and only one of the following is true:
   (i) a = b;
   (ii) there exists another length c, such that a = b + c;
   (iii) there exists another length d, such that b = a + d.
2. For any lengths a and b, a + b > a.
3. For any lengths a and b, a + b = b + a.
4. For any lengths a, b, and c, a + (b + c) = (a + b) + c.
5. For any length a, there is another length, b, such that b < a.
6. For any lengths a and b there is another length, c, such that c = a + b.
7. For every nonempty class of lengths having an upper bound, there is a least upper bound.

(Michell 1981, 215)

While these conditions suggest "length" in their specifications, the sticks "produce" length—a considerable difference inasmuch as specification denotes a retrospective mathematical process by means of axioms, whereas the sticks fabricate length as a measure appearing before our eyes. The value of axiomatic specification is to provide a succinct, internally consistent process of logical steps, but as Guttman wrote, "Even in mathematics, axiomatization is an intermediate developmental stage; one must first have some idea of some body of inter-relationships for which the axiomatization may be fruitful" (1971, 346). The ideals of exactness and rigor in mathematics are the product of time and refinement. The order of sticks manufactures length from operations that evolved early and developmentally; this is the same process that might initially produce any variable. Instrument refinement requires theory, continuous quality control, validation, and so on—but that is another story!

Our felt experience of "hot" and "cold" is crudely qualitative. Measures of temperature are made quantitative by fashioning a graduated tube constructed of glass together with mercury contained in a vacuum. This approach follows from numerous experiments using water, alcohol, and so forth. The instrumentation improved because the goal remained constant. Today, we have achieved a uniform association between the expansion of mercury and a measure of temperature. The sensations of cold and heat are derived from the human organism, while temperature is manufactured analogously by an instrument (a thermometer) in a process resembling the use of sticks to manufacture length. Wittgenstein (1958, 508) writes, "I am not used to measuring temperatures on the Fahrenheit scale. Hence, such a measure of temperature 'says' nothing to me." To those who are familiar with the scale, F-70 means pleasant and F-32 means freezing. The matter rests upon an analogy, and the associations that are important for interpreting a measure. A NexTemp (2004) thermometer is strikingly similar to the sticks analogy but uses "chemical cavities" calibrated by a chemical specification equation instead of sticks to box in one's temperature.

Temperature is constructed by employing comparison and order, while sensation usually lacks clear lines of demarcation further hampered by the "swamp of language." Thermometers employ a sequence of units (numbered) which, when correctly constructed and interpreted, produce an unambiguous result. A sensation lacks specificity, while a working thermometer produces a consistent, useful value. When Chang ([2004] 2007) selected *Inventing Temperature* as the title of his book, he proclaimed the essence of the history of measurement: measuring is invention, the process of variable construction, and that inventive process is never ending.

Considering the distinction between reality and the idealized goal, we might ponder G. H. Hardy's remark that "nothing practical" would occur in his *Course of Pure Mathematics* (1908), except "constructing" a world of mathematics.

We dwell in this fabricated land, straddling between two realms—one real and the other ideal.

6

COMPARISON IS KEY

Rasch (1977) addressed science and objectivity in his last published paper, *On Specific Objectivity: An Attempt at Formalizing the Request for Generality and Validity of Scientific Statements*. In section III (68), he rhetorically asked, "What is Science?" His answer specified two conditions. Science is:

1. Making comparisons, and
2. Making these comparisons objectively.

Rasch (1977) further explained these two conditions:

Two features seem indispensible in scientific statements. They deal with comparisons, and the comparisons should be objective. However, to complete these requirements I have to *specify the kind of comparisons and the precise meaning of objectivity.* (68)

Rasch (1977) follows the above quote by another one also taken from an earlier paper (Rasch 1967) but with each part given important headings:

Specifying Comparisons. Consider a class of 'objects' to be mutually compared. The sense in which they should be compared is specified through a class of 'agents,' to each of which each object may be 'exposed.' On each exposure an 'observation' – quantitative or qualitative – is made. The whole set of such observations made when a finite number of objects $O_1 \ldots O_n$ are exposed to a finite number of agents $A_1 \ldots A_k$ from the data from which comparisons of the O's ... to such agents as the A's can be inferred reactions.

Specifying Objectivity. Now, within this framework, which I have taken from psychophysics, the 'objectivity' of a comparative statement on, say two objects, O_1 and O_2 is taken to mean that although being based upon the whole matrix of data it should be independent of which set of agents $A_1 \ldots A_k$ out of the available class were actually used for the comparative purposes, and also of which objects … other than O_1 and O_2 were also exposed to the set of agents chosen.

Specific objectivity, some general properties. In order to distinguish this type of objectivity from other uses of the same word I shall call it *'specific objectivity,'* and in passing I beg you notice the relativity of this concept: it *refers only to the framework specified by the class of objects, the class of agents and the kind of observations which define the comparison.* (2–3)

Rasch (1967) adds this important qualification:

> … the objects and/or agents are subject to comparison … the data themselves are not directly compared, they only serve as the instruments for the comparison aimed at. The consequences of introducing these two concepts: (specific) comparisons and specific objectivity, completed by the requirements that a comparison is always possible and its result always unambiguous, are really overwhelming. (2–3)

The quotations above serve as an introduction to what is required in the method of comparison as well as the consequences of making such comparisons. It is not the data but what it stands for that is the goal of science. The essential conditions are two: (1) "specify the kind of comparisons," and (2) specify "the precise meaning of objectivity."

These stipulations suggest the following paradigm:

Propositions:	Specifications:
1. Scientific statements	by the kind of comparisons made.
2. Objective comparisons	by the precise meaning and achievement of specific objectivity.

Scientific statements are the consequences of specifying comparisons. Objective comparisons are the consequences emanating from comparisons exemplifying the precise meaning of specific objectivity. Rasch (1977) draws attention to the ubiquity of observations:

> ... comparisons form an essential part of our recognition of
> our surroundings: we are ceaselessly faced with different
> possibilities for action, among which we have to choose
> just one, a choice that requires that we compare them. This
> holds both in everyday life and in scientific studies. (68)

These remarks appear general, obvious, and rather simple. Every day finds us comparing prices and products or deciding activities. Rasch's observations on making comparisons in everyday life or pursuing science are self-evident but clearly essential. However, a careful examination of the exact role for making comparisons is required.

Rasch (1977) next responds to those who suggest measurement as primary in conducting science: "That science should require observations to be measurable quantities is a mistake, of course; even in physics observations may be qualitative – as in the last analysis they always are" (2). This statement is vitally important in the social sciences because it challenges two common beliefs. First, Rasch dismisses the notion of measurement as a prerequisite for science to proceed. His quote echoes Louis Guttman's comments of a similar nature:

> I have avoided the term 'measurement' in all my
> writings and teachings. I have found it neither useful
> nor necessary ... No fixed a priori collection of abstract,
> contentless techniques or principles can be universally
> appropriate for scientific progress. (330–31)

Second, Rasch raises qualitative observations to the prominence rightly deserved. Guttman shared this view also, saying, "The basic data of most mental tests are qualitative, yet no treatment is given of the theory of such qualitative data" (Levy 1994, 324). Guttman (1971) argued that his approach to science is via "... hypothesis construction for aspects of a universe of observations recorded" (333). He eschewed measurement while promoting better investigations driven by theory. Guttman's remarks further suggest interesting comparisons between these two iconoclasts of traditional measurement practice. Andrich (1982, 1985, 1988) has made psychometric comparisons of Guttman and Rasch perspectives. Englehard (2008) has examined both perspectives relative to invariant measurement.

A contemporary expression of many of these same issues has been given by Rein Taagepera (2008), whose book *Making Social Sciences More Scientific* pursues these same goals and others in refocusing attention on constructing predictive models. The need to repeat calls to the goal of measuring as predicting serves to show how important these matters are but also how neglected in practice they have been.

The important issue developed from Rasch thus far is that only specific comparisons, grounded in substance/objects, produce generalities and laws. It is the properties of these generalities and laws that we employ for guidance and direction. Data and its analyses are only a means to this end and should not be mistaken for the conclusion. In practice, this means that what is derived from data must reach beyond to become a generalized outcome. Whether it does or not is critical.

The Strategy of Comparison

Comparison is driven by theory and experiment. Basic statements about making comparisons may initially appear too simplistic. What makes these elementary concerns worthy of further investigation? Why such attention to making comparisons? Rasch illustrates his points with two examples. The first example involves a comparison of ashtrays. His choice of this example is unfortunate for the times but illustrative nevertheless. Data from this experiment is produced by dropping heavy and light ashtrays from six different heights. It is reported as follows:

Falling Distance

	H_1	H_2	H_3	H_4	H_5	H_6
Heavy ashtray	+	+	+	+	-	-
Light ashtray	+	+	-	-	-	-

Survival intact is denoted +, and breakage is denoted -. While the experiment is simple, Rasch reminds us that qualitative comparisons are fundamental to science. He stresses the importance of recognizing qualitative methodology as indigenous to science. This example, like any example, requires confronting each object with an action to produce a reaction—a tripartite condition. But objects and actions must be systematically arranged to allow a valid observation to be made. A theory is stimulated by a thought experiment, an intuitive insight, or a historical review of relevant literature.

The experiment consists of a tripartite condition of *object, agent,* and *resulting reaction.* In a psychometric application, this could be person, item, and response. Rasch describes the process:

1. To determine whether an object has a certain property one must do something to the object, confront it with some action or different actions liable to create one of a number of reactions.
2. If knowledge gained in this way is to be used in making a choice it must be obtained for several objects of the kind in question so that a comparison becomes possible. (69)

Rasch (1977) introduces his second example showing the sequential development of gas equations (IV:1–IV:15) not discussed here. The essential points that Rasch draws from the more rigorous exposition of this second example (71–73) are the following:

1. Varying the two parameters of temperature and pressure introduces a stepwise series of systematic experiments,
2. The results show "linearity to be a pervading trait,"
3. This results in "a law of greater generality,"
4. The sequence progresses from observations of volume, pressure and temperature and passes through a series of stepwise comparisons leading to summary equations.

Rasch concluded:

> This procedure, it seems, can be taken as a prototype of the experimental charting of a complex field ... only through systematic comparisons – experimental or observational – is it possible to formulate empirical laws of sufficient generality to be – speaking frankly – of real value, whether for furthering theoretical knowledge or for practical purposes. ... I see *systematic comparisons as a central tool* in our investigation of the outer world. (74)

Rasch gives great importance to building a sequence of experiments. It is not a singular, critical experiment that typifies science although it has occurred at times but a series of them, each of which builds upon the other. Parameters may be experimentally manipulated to observe outcomes and confirm or deny theoretical predictions. The way that Rasch demonstrates development of the gas law brings a theory to an encompassing state built from and supported by a succession of critical experiments whose hypotheses are sequential, stepwise, and linear. This process constitutes Rasch's *causal model*. We have earlier stipulated, "The Rasch model in concert with a substantive theory is a powerful tool for discovering and testing the adequacy of formulations" (Burdick, Stone, and Stenner 2006, 1059), and specified, "It takes a substantive theory to unambiguously distinguish between a latent variable and an index" (Stenner, Birdick, and Stone 2008, 1177). Rasch was clearly proposing that comparisons be made in the context of substantive theory driving experimental studies. Only in this context can claims against latent variables be made following the specification outlined above.

In *Comparisons and Specific Objectivity*, Rasch (1977) generalizes and formalizes the implications of his two examples (75). He begins with explanations and equations more rigorously defining comparison and specific objectivity that are abstracted below but sequenced for order:

... two collections of elements O and A denoted here objects and agents ... single elements and indices O_v and A_i ... enter into a well-defined contact C ... every contact an outcome R.

... the three collections of elements ... the frame of reference
$$F = [O, A, R] \qquad\qquad 1$$

the concept of comparison ... contact C determines outcome R as a function of object O and agent A
$$R = r (O, A). \qquad\qquad 2$$

Within a specified determinate frame of reference a comparison between two objects O1 and O2 – with regard to their reactions to containing the agent A – is defined as a statement about them which is based solely on those reactions.
$$R_1 = r (O_1, A), R_2 = r (O_2, A) \qquad\qquad 3$$

... this comparing function u (R1, R2) forms a collection U
... the elements ... may be qualitative
... inserting function U in

$$U (R_1, R_2) = u (r (O_1, A), r (O_2, A)) \qquad\qquad 4$$

and a statement about O1 and O2 ... will depend on A, the agent.

the comparing function U(R1, R2) as a function of O_1 and O_2 conditioned by A

$$U (r (O_1, A) r (O_2, A)) = u (O_1, O_2 \mid A) \qquad \qquad 5$$

as a comparator for O_1 and O_2 conditioned by the agent A (in analogy to the concept of conditional probability).

Statements dependent on the agent [object] are said by Rasch to be local comparisons.

Clarity is given about what is and what is not objective, what is local or specific in the frame of reference:

> Local comparisons are ... useful as pointers ... a comparing statement ... O_1 and O_2 to be more than locally valid, the comparator must be independent of which A from A has been used to produce the reaction ... if the condition is fulfilled ... denote the comparison between these two objects as for agents between these two objects

$$U (R_1, R_2) = u (r (O_1, A), r (O_2, A)) u (O_1, O_2). \qquad \qquad 6$$

> ... if this globality within A holds for any two objects O_1 and O_2 in O ... the pairwise comparison as defined by V:4 as specifically objective within frame of reference F.

> The term 'objectivity' refers to the fact that the result of any comparison of two objects within O is independent of the choice of the agent a_i within A and also of the other elements in the collection of objects O; in other words: independent of everything else within the frame of reference, than the two objects which are to be compared and their observed reactions ... the qualification "specific" is added because the objectivity of these comparisons is restricted to the frame of reference F in [1] ... denoted as the frame of reference for the specifically objective comparisons in question. (75–77)

Rasch makes very clear:

> ... specific objectivity is not an absolute concept, it is related to the specific frame of reference ... this definition

concerns only comparisons of objects, but within the same frame of reference it can be applied to comparisons of agents as well. (77)

The philosophic issues encompassing the history of objectivity appear circumspectly avoided in these statements. Rasch specifies: "The concept has therefore not been carved out in a conceptual analysis, but on the contrary its necessity has appeared in my practical [statistical] activity" (58).

Why avoid the philosophic issue of causation? Speculation suggests it was not in good taste at the time he was writing to speak of causation in general. The zeitgeist did not appear to support objectivity and causality in this context. The consequences of quantum mechanics and the influence of logical positivism may have kept Rasch from wanting his methods contaminated by any digression into philosophy, or to have metaphysical issues injected into his systematic discourse. The strategy seems very clear in light of the quotes given in the opening paragraphs of this paper. Rasch wants to make his case without contending with excessive philosophical baggage concerning causality and objectivity. To illustrate the zeitgeist Rasch was avoiding, consider the opening sentence to Waismann's chapter entitled "The Decline and Fall of Causality," chapter V in *Turning Points in Physics* (1961): "1927 is a landmark in the evolution of physics – the year which saw the obsequies of the notion of causality" (84).

Hoover (2004) offers another illustration:

[producing] ... a dip of about twenty percent in the occurrence of the causal family from the 1950s [for 'causally', 'causality', or 'causation' in econometric literature]. (152)

Rasch is by no means conservative regarding the promotion of objectivity via theory/experiments constructed with a view to exploring results based upon engineered methods. He employs the word "law" more than a dozen times throughout his 1977 paper. While Rasch does not venture into "causation" stated explicitly, he does so implicitly via employment of comparison as described above. It seems clear, except for evading the philosophic realm of causality, Rasch advocates a causal mode of investigation by means of comparisons founded upon hypotheses and data.

It is important to observe his distinction between "indicators" and "specific objectivity." Rasch (1977) specifies,

> ... if this globality within A holds for any two objects O_1 and O_2 in O ... the pairwise comparison is defined by (4) as specifically objective within the frame of reference F.

> The term 'objectivity' refers to the fact that the result of any comparison of two objects within O is independent of the choice of the agent a_1 within A and also of the other elements in the collection of objects O; in other words: independent of everything else within the frame of reference, than the two objects which are to be compared and their observed reactions. (76)

The essential point rests upon the comparison of two objects (or agents) independent of agent (or object) in the collection of objects (or agents) and their observed reactions within a specified frame of reference. As indicated earlier in the quotes above, not all comparisons meet the conditions of specific objectivity. A key issue is distinguishing "... those statements dependent on the agent (object)," specified by Rasch to be "local comparisons" from those that emanate from "specific objectivity."

Returning to Rasch's ashtray example, we find there are no differences observed for the two ashtrays dropped from heights H_1 and H_2 because both survive breakage. No difference results from heights H_5 and H_6 also because neither ashtray survives the fall. The first two heights, H_1 and H_2, allow no comparison to be made, and the same occurs from observing H_5 and H_6. But every result occurs from making comparisons without knowledge of the heights or the composition of the ashtrays. This is a subtle but critical point in Rasch's methodology of comparisons. Nothing is required beyond the gross descriptions of ashtray composition/construction and a sequence of heights for these drops. Rasch (1977) concludes:

> Comparing ashtrays at H_3 and H_4 shows the two middle distances with the heavier ones surviving breakage while the lighter ones do not ... objects type – 2 elements (heavy, light); agent distances – 6 elements (each once ordered above the other); and reaction elements – 2 results (survived breakage, did not survive breakage). (69)

These specified elements reiterate Rasch's earlier contention that it is qualities that are engaged, while measurement quantities are not required.

> What is required, ... is to do something specific to the object, confront it with some action or different actions

liable to create one of a number of actions. If knowledge gained in this way is to be used in making a choice it must be obtained for several objects of the same kind in question so that a comparison becomes possible. (69)

Rasch (1977) draws these inferences from his ashtray experiment:

A possible comparing function could be the assertion 'No.1 is more solid than no.2,' defined operationally by the sequence of reactions + -, that is, the first one holds and the second one breaks. This comparison is not global, it has the value 'true' for the intermediate falling distances and 'false' for the other. (77)

Another comparing function is 'no 1 is at least as solid as no. 2,' defined operationally by the observed reactions + + or + - : either they both hold or they both break or only no 1 holds. This comparison is global within the frame of reference of the described experiment and can even be expected to be global also if more ashtrays and more falling distances are included in the frame of reference. (78)

This brings about the ashtray conclusions:

1. Ashtray No. 1 is not more solid that No. 2 across all heights. It is not global because it is true for some heights but not all. A general statement cannot be made for the two types of ashtrays over all heights.
2. Ashtray No. 1 is as solid as No. 2 can be supported as global in that No. 1 (heavy) is equal to or exceeds No. 2 (light) in heights 1 to 4, although both break at heights 5 and 6.

Rasch next delineates specific objectivity as separate from consideration of general objectivity.

... specific objectivity is not to be expected from an arbitrary chosen comparing function of $u(R_1, R_2)$ bifactorial frames of reference" [where] every reaction is characterized by a so-called scalar parameter ... characteristic of object, agent or reaction.

... parameters O_v, A_i and $R_{vi...}$ denoted ω_v, α_i, ξv_i.

… the reaction is assumed to be uniquely determined by object and agent …

$$\xi_{vi} = \varrho \, (\omega_{v}, \alpha_{i})$$

VI:1

the parametric reaction function.

The condition corresponding to V:6 for specific objectivity of comparison of objects …

ω_{λ} and ω_{v} is

$$\upsilon \, (\varrho \, (\omega_{\lambda}, \alpha), \varrho \, (\omega_{v}, \alpha)) = \upsilon \, (\omega_{\lambda}, \omega_{v}).$$

Under these conditions a decisive statement can be made on the properties of the reaction function o (w, a) that are necessary for establishing specific objectivity of comparisons of objects within the framework F (77–78).

Rasch advocates strategies of comparison designed to tease out knowledge. His first example builds the case. The outcomes from the ashtray experiment can be arranged in a two-by-two table as follows:

		Heavy	
		Breaks	Survives
Light	Survives	no data	H_1, H_2
	Breaks	H_5, H_6	H_3, H_4

We already know enough about ashtrays and height to dismiss this simple example as unnecessary, but then we would miss the point of his generalizations that follow. Rasch's experiment establishes these major points.

1. Systematic comparisons made under specified conditions produce stochastically consistent results.

2. It becomes insightful science to systematically arrange each encounter between an object and an agent and then observe the result, the classic experiment for destructive testing.

3. Given enough such experiments, wisely contrived, we can often predict outcomes whenever we gain an understanding of the matter under study and our theory is sound. But sometimes our predictions are surprising and wrong.

There is a clear difference between the heavy and light ashtrays from the results for H_3 and H_4 because the heavy one, H_3, survives, and the light one, H_4, does not. This establishes Rasch's critical point illustrated by this experiment: comparisons are fundamental to measurement. Rasch also reminds us the results were produced by qualitative observations, and he indicates this is frequently the case with many scientific investigations in the physical sciences. There has been no requirement or need for prior quantitative measures in order to produce these results. We might wish to bring further clarification and more sophistication to this experiment by refining the conditions (introducing different heights with different ashtrays) and observing the results, but specific units for height and mass are not required. Order suffices.

Constructing Experimental Comparisons

Rasch argues that the process and strategy of constructing comparisons is the essence of scientific methodology. In his ashtray experiment, every object comes in contact with every agent via a frame of reference. Each interaction produces some outcome resulting from this intersecting occurrence of agents with objects. Outcomes are recorded by one of two qualitative values in a dichotomous frame of reference defined by ashtrays and heights.

A two-by-two frame of reference permits these comparisons to be ordered in a systematic way. The outcome may be qualitatively the same or different, as in the ashtray example. Further comparisons are made possible by progressing through a larger data frame and subsequently aggregating all such useful comparisons regarding height and ashtrays. Order remains fundamental regardless of how complex the comparison framework grows.

He characterizes pair-wise comparisons of objects (or agents) as *"specifically objective within the frame of reference"* (77).

Rasch (1977) delineates the process:

> The term 'objectivity' refers to the fact that the result of any comparison of two objects within O is independent of the choice of the agent A within A and also of the other elements in the collection of objects O, in other words: independent of everything else within the frame of reference than the two objects which are to be compared and their observed reactions. And the qualification 'specific' is added because the objectivity of these comparisons is restricted to the frame of reference F. This is therefore denoted as the frame of reference for the specifically objective comparisons in question. This also makes clear that the specific objectivity is not an absolute concept, it is related to the specified frame of reference. (77)

> Designating ω_{vl}, α_l, and ξ_{vl} as parameters for O_v, A_l, and R_{vl} gives

> $\xi_{vl} = (\omega_v, \alpha_l)$ the parametric reaction function.

> This condition for the specific objectivity of comparisons for objects ω_λ and ω_v is

> $\upsilon \, (\varrho \, (\omega_\lambda, \alpha), \varrho \, (\omega_v, \alpha)) = \upsilon \, (\omega_\lambda, \omega_v)$.

Rasch then formulates his main theorem of specific objectivity:

> Let objects and agents in the bifactorial determinate frame of reference F be

> characterizable by scalar parameters ω and α and reactions by a scalar reaction function of 'convenient' mathematical properties

> $\xi = \varrho \, (\omega, \alpha)$,

with three monotonic functions:

> $\acute{\omega} = \phi \, (\omega), \acute{\alpha} = \psi \, (\alpha), \acute{\xi} = \chi \, (\xi),$

transforming the scalar reaction function to an additive one:

$$\xi = \acute{\omega} + \acute{\alpha},$$

... a necessary and sufficient condition for specifically objective comparability of objects as well as of agents. (79)

The model becomes the source of objective measurement and not the details of the data. Hence, the sometimes used phrase "When data fit the Rasch model ..." might be better expressed "The Rasch model has (1) identified a fit of data to the model across a frame of reference implied by this experiment, or (2) identified a lack of fit between the same."

Rasch (1960) had earlier alluded to this same condition:

It is tempting, therefore, in the cases with deviations of one sort or other to ask *whether it is the model or the test that has gone wrong* ... the question is meaningful ... the applicability of the model must have *something to do with the construction of the* test. (51)

The model confirms or identifies inconsistencies by making experimental comparisons of agents, objects, and outcomes from the data. Inconsistent comparisons in an experiment suggest theory, data, or both are suspect. Rasch's penchant for constructing data plots as a "check on the model" indicates his awareness of the importance of quality control. We further draw attention to the role of construction or engineering as critical. Every experiment is a fabrication in the sense of manufacturing a desired outcome from theory and data. To the degree that we succeed, we know what is required to produce the desired outcome. Failure or deviations indicate full knowledge is lacking.

In such instances, theory, data, or both require further investigation. This is an important point for applying the model to data. To use an illustration from physics, as Rasch often did, Newton's law of motion rests upon his model and not on data. It is the law and not the data to which we attend. Furthermore, we recognize data is fraught with contamination from many sources of error. Hence, we generalize from Newton's model of force/mass—his abstraction produced from contaminated data. Continuous validation has supported his theory until the advent of subatomic physics required a new viewpoint, but even this evolutionary change has not obviated Newton's law.

Rasch's example began with six ashtrays, but only two ashtrays provided unique and key information. With a supply of additional ashtrays made of different types (shapes, composition, etc.), we can proceed to make every two-way comparison of different ashtrays dropped from various heights until we have exhausted all the two-way (height by ashtray) comparisons useful for this crash test. Summarizing the findings provides types of ashtrays arranged by their capacity to withstand breakage according to height employed. From a comparison of two similar ashtrays from different heights (or two different ones from the same height), this simple comparison may be extended as far as desired. We do not need to physically order the ashtrays, although we could do so, because a record of success and failure is sufficient. A summary of all the two-by-two comparisons will produce an ordered arrangement of all the ashtrays by all the heights employed. This process results in a durability/survival variable, identifying ashtrays from the most fragile to the most durable, and from the lowest to highest heights in the frame of reference deemed useful for the experiment. We can also move from a two-way comparison to a multivariable frame if desired.

Any expansion embodies the essence of experimentation guided by theory. It produces an ever-encompassing frame of reference for determining durability for a variety of ashtrays dropped from a variety of heights. We can confirm previous predictions from theory, as well as make further predictions about unexamined ashtrays and unexamined heights. As we proceed, it may be possible to derive other predictive hypotheses regarding what is expected to occur. A theory of ashtray composition dropped from various heights will probably produce additional predictive insight. Outcomes may be increasingly predictive compared to what was known initially. Theory regarding ashtray composition and heights may become increasingly better understood. Essential to the process is confirmation by cross-validation. If desired, we can describe the order with numerals noting once more that measures follow comparison.

Campbell ([1921] 1953) offers a relevant remark:

> If measurement is really to mean anything, there must be some important resemblance between the property measured, on the one hand, and the numerals assigned to represent it on the other hand. In fundamental measurement this resemblance (or the most important part of it) arises from the fact that the property that is susceptible

to addition following the same rules as that of numbers. There is left resemblance in respect of "order." ... Order then is characteristic of numerals; it is also characteristic of the properties represented by numerals. This is the feature which makes "measurement" significant. (126–127)

The key to measurement in psychometrics is (1) making systematic comparisons (items, persons, judges, etc.) and (2) making these comparisons objectively. Order becomes paramount. Assigning numerals in such cases is really an afterthought stemming from the initial pair-wise comparisons of ashtrays to systematize findings and quantify the results. The numerals/numbers assigned and utilized become the summary of the experimental findings. They follow the experiment but do not dictate it. Only when the results are valid and confirmed can we successfully apply/substantiate numeric/algebraic abstractions, and not the other way around. We construct measurement from results of the theory/experiment. We have Celsius and Fahrenheit scales but one and the same temperature at a simultaneous measurement.

In Rasch's words (1977),

> Objectivity is achieved when a comparison of any two objects is independent of everything else within the frame of reference other than the two objects which are to be compared and their observed reactions. (77)

Rasch (1977) also shares his emphasis upon comparison with the English philosopher Hume ([1817] 1949), who wrote:

> All kinds of reasoning consist in nothing but a comparison, and a discovery of those relations, either constant or inconstant, which two or more objects bear to each other. (77)

Making systematic comparisons constitutes the fundamental process by which we determine essential differences. Systematic comparisons employing transitivity produce order. Measures considered to be quantitative actually result from qualitative ordering. Comparison first makes clear how any two objects/agents relate to each other by whether one is "more" than the other. This is the grounding for all measurement. Hume stated the importance of comparison as a logico-philosophical deductive principle. Rasch specified it using a simple inductive example to deliver a mathematical generalization. Generalization from order via comparison produces a ground to measurement. Systematic comparisons produce order and understanding without the necessity of any a priori measurement scheme. Ricoeur (1977) speaks to their value and power:

> … the power of making things visible, alive, actual is inseparable from either a logical relation of proportion or a comparison … Thus one and the same strategy of discourse puts into play the logical force of analogy and of comparison – the power to set things before the eyes, the power to speak of the inanimate as if alive, ultimately the capacity to signify active reality. (34–35)

The two-way frame is the data structure we use to determine order from the results observed by systematic comparisons produced from the tripartite elements of agent, object, and outcome. Comparisons derived from this two-way frame are objective whenever "they are independent of everything else within the frame of reference." We restate this principle by some alternate expressions:

1. The frame of reference provides the basis for making objective comparisons.
2. Comparisons produce order in a frame of reference guided by theory.
3. Order (for agents, objects, and results) is demonstrable (or not) within the frame of reference.
4. Comparisons are specifically objective when made within the frame of reference (2 and 3).

In a deterministic framework, the consequences of comparison remain categorical (Guttman). In a stochastic framework, the outcomes of comparisons are probabilistic (Rasch). Not every drop of a specified ashtray from a specified height produces the exact same outcome. There will be a distribution of error surrounding each comparing event. But if the experimental conditions are the same, then a probabilistic result provides the answer. Residual analysis becomes the important tool by which to evaluate every outcome resulting from comparisons. Rasch did not address quality control directly in his 1977 paper, but he did give careful consideration to always confirming the model to data as shown by his constant attention to plots and graphs. He relied on data plots, not correlations, to show confirmation or identify deviance. This is good advice for today as well.

The conclusions to be drawn from Rasch's two examples are as follows:

1. We construct science by making comparisons. These comparisons must be made by following a procedure leading to specific objectivity. Theory guides this process, but experimentation confronts the outcome of hypotheses with data. This strategy assures clarity in the process and allows replications to confirm or refute the results.
2. The two-way frame of reference specifies the agent, object, and resultant outcomes in a predictive guise.
3. The two-way frame of reference arranges and subsequently summarizes the comparisons. These comparisons are fundamental to what Rasch designated as specific objectivity.
4. Measurement follows from the results of qualitative comparisons that have been constructed in a systematic way, using order as the fundamental characteristic.

Additive Conjoint Measurement

Rasch's experimental conditions demonstrate the first of two specifications essential for additivity—that is, the rows and columns of a data matrix are monotonic for the data (Krantz, Luce, Suppes, and Tversky 1971). The independent variables of ashtrays—composition, height, and result— were ordered in the ashtray experiment by a two-way frame of reference. Continued experimentation could sustain and expand upon the range by expanding the frame of reference through additional experimentation.

A second property, double cancellation, identifies departures from additivity (Krantz, Luce, Suppes, and Tversky 1971). Luce and Tukey (1964, 3) show a simple way to graph the effects of two factors for demonstrating this property. Borsboom (2005) sees double cancellation as the consequence of additivity that "brings out the similarity between additive conjoint measurement and the Rasch model." He indicates that the condition of independence is "similar to what Rasch called parameter separation" (97).

Perline, Wright, and Wainer (1979) presented data to argue Rasch's psychometric model as a special case of additive conjoint measurement. Additivity is ascertained by examining these two necessary ordinal properties. Coombs, Dawes, and Tversky (1970) declare all independent variables are measured on an interval scale with a common unit when additivity exists.

The essential point of Perline, Wright, and Wainer (1979) in their analysis of two studies was that the Rasch model estimates (when data is orderly) correspond to the conditions for additive conjoint measurement, but data fit does not have to be ideal to be demonstrable.

Borsboom (2005) argues, "… the Rasch model has little to do with fundamental measurement, In fact, the only things that conjoint measurement and Rasch models have in common, in this interpretation is additivity" (132). Indirectly, this simply confirms the comparability of Rasch estimates of person and item parameters to additive conjoint measurement. The key property of additive conjoint measurement rests upon order determined from making comparisons, which returns us once more to the points made from Rasch's ashtray example.

Newby, Conner, Grant, and Bunderson (2002) provide a clear statement of the mathematical relationship between additive conjoint measurement and the Rasch model, saying, "… it is not the case that, given any data, the Rasch model will provide a natural numerical representation of the data that is comparable with ACM" (350).

Why should the Rasch model address data except to show if the conclusions drawn from making constructive comparisons are justified or not within the frame of reference? Our point is that a so-called Rasch model analysis does not cleanse data. Nor does it signify the conclusion of data analysis. Further analysis must be conducted in the context of theory with clearly stated hypotheses. Removed from the context of theory, no analysis sui generis provides substantive answers except incidentally (Stenner, Burdick, and Stone 2008). For example, no statistical transformation of so-called raw scores to IQ values and so on makes any improvement to a substantive understanding of intelligence. Comparisons are the experimental results/consequence of an interaction of agents and objects guided by theory/hypotheses (Brogdin 1977). Comparison remains the key to any investigation. Content and substance are embodied in the selected comparisons made.

Are the results obtained from any Rasch analyses descriptive, or are they predictive? Does the ashtray frame of reference only describe, or can it predict? Does what we learn from dropping a variety of ashtrays from a variety of heights yield merely descriptive observations? Can these results predict stochastically suggested outcomes? Rasch answered these questions by the quotes we provided earlier.

Concerning parole data from Perline, Wright and Wainer (1979), we ask, can causality can be ascribed to the outcome emanating from this Rasch frame of reference? What do we learn from the parole data that provides a prediction? Is it a truly causal matter? Can we ascertain any information from the experimental manipulation of the tripartite object, agent, and outcome in a frame of reference to produce expected outcomes thought to be causal? If so, the variable moves beyond mere description to prediction and the consequences move to a higher plane (Stenner, Stone, and Burdick 2008). But how do we truly ascertain prediction and causality in Rasch measurement?

Our answer suggests the typical Rasch analysis is largely descriptive and lacks evidence about a causal connection or experimental manipulation. A Rasch analysis addresses order in the data. Such analyses may describe association or correlation, but they are not instances of causal inference—suggestive, maybe, but not confirmatory. Lack of evidence suggests a crucial next step because as Stenner, Burdick, and Stone argue (2008), "There is no single piece of evidence more important to a construct's definition than the causal relationship between the construct and it indicators" (1153).

With respect to the parole data, we ask how these nine items might be placed in an experimental framework so as to examine their usefulness and predictive validity. The answer is to put these comparisons in a context for testing causality. The Rubin-Holland (Holland 1986; Holland 1980) framework for demonstrating causal inference specifies *no causation without manipulation*. Rasch analysis has produced considerable understanding of the variables exemplified in the resulting map of items, persons, and results from the two-way framework in his example. The result remains descriptive unless contained within an experimental context guided by theory. Supporting any experiment should be a theory guiding inferences about what produced these item outcomes. In the matter of parole, for example, how is granting or not granting parole to be experimentally controlled? How will outcomes be determined? The hallmark of success is formulation of a specification equation to predict and validate successful manipulation of the variables required in a probabilistic causal model portrayed in the measuring variable (Pearl [2000] 2009).

We therefore specify a process model for classifying Rasch investigations:

1. Description	2. Explanation	3. Causation
Association	Prediction	Physical data giving substantive evidence over a
Correlation	Theory/Equation	wide range of changes based upon theory.

1. The comparison process begins with an idea, a conjecture, or history, which develops into a thought experiment.
2. The experiment is designed and conducted:

   Input $\longrightarrow$ Manipulation $\longrightarrow$ Output
3. An equation may be formulated and specified:

   Domain $\longrightarrow$ Function $\longrightarrow$ Range
4. Plot(s) of data and analysis of effects and residuals provide information and data/model control.
5. From such strategies come the properties of comparison and order guided by theory, the essential ingredients for achieving specific objectivity.

Box 17: Steps from Quality to Quantity

Too often the calibration function is mentioned but not described in detail, which is especially frustrating for someone newly introduced to Rasch measurement procedures. These steps inform the reader about the sequence followed to transform counts correct to calibrated values. More importantly, we describe the continuing steps to be pursued beyond merely fitting data to a Rasch model.

We start with counts such as the number of correctly spelled words (a count) on a weekly spelling test. Additional figures give a more detailed example, followed by further steps that are necessary to complete the process. A matrix of counts by items for the entire class gives rise to the data matrix by which to make summations of the item and person scores or counts for each member of the class—a frame of reference. Counts correct or percentages are useful in the classroom, but measuring spelling ability is another matter

Rasch's logistic model [1] is utilized by which to compute parameter estimates of the item difficulties and person measures from their score/count to a probability.

$$P_{nix} = [1] \quad \frac{\exp [x_{ni} (B_n - D_i)]}{1 + \exp (B_n - D_i)}$$

In equation [1] n = person ability B (or b) in spelling for this example, and i = item (spelling) difficulty D (or d) for each spelling word, giving the probability for each cell in the matrix. This results from application of equation [2], a causal Rasch model for dichotomous data, which in this spelling example sets a measurement outcome (expected score) equal to a sum of modeled probabilities:

$$Expected\ score =: \sum_i \frac{e^{(b-di)}}{1 + e^{(b-di)}}$$

(2)

The measure (e.g., person parameter, b) and the parameters d_i pertaining to the difficulty d of item i) are independent variables in this frame of reference. (An item parameter d_i and the parameters b_v for persons could also be computed.) The measurement outcome (e.g., count correct on a spelling test) is observed, whereas the person measures and instrument parameters are not observed but can be estimated from the response data matrix.

Score Group	Items Ability	1 -6.75	2 -6.38	3 -4.89	4 -4.86	5 -3.63	6 -3.51	7 -2.60	8 -2.38	9 -2.07	10 -1.55	11 -1.98
0	-8.85	-2	-2	-4	-4	-5	-5	-6	-6	-7	-7	-7
1	-7.37	-1	-1	-2	-3	-4	-4	-5	-5	-5	-5	-6
2	-6.25	1	0	-1	-1	-3	-3	-4	-4	-4	-4	-5
3	-5.39	1	1	-1	-1	-2	-2	-3	-3	-3	-3	-4
4	-4.67	2	2	0	0	-1	-1	-2	-2	-3	-3	-3
5	-4.04	3	2	1	1	0	-1	-1	-2	-2	-2	-2
6	-3.47	3	3	1	1	0	0	-1	-1	-1	-1	-2
7	-2.94	4	3	2	2	1	1	0	-1	-1	-1	-1
8	-2.44	4	4	2	2	1	1	0	0	0	0	-1
9	-1.95	5	4	3	3	2	2	1	0	0	0	0
10	-1.45	5	5	3	3	2	2	1	1	1	1	0
11	-0.93	6	5	4	4	3	3	2	1	1	1	1
12	-0.41	6	6	4	4	3	3	2	2	2	2	1
13	0.13	7	7	5	5	4	4	3	3	2	2	2
14	0.65	7	7	6	6	4	4	3	3	3	3	2
15	1.16	8	8	6	6	5	5	4	4	3	3	3
16	1.66	8	8	7	7	5	5	4	4	4	4	3
17	2.16	9	9	7	7	6	6	5	5	4	4	4
18	2.64	9	9	8	8	6	6	5	5	5	5	4
19	3.13	10	10	8	8	7	7	6	6	5	5	5
20	3.62	10	10	9	8	7	7	6	6	6	6	5
21	4.12	11	11	9	9	8	8	7	7	6	6	6
22	4.63	11	11	10	9	8	8	7	7	7	7	6
23	5.18	12	12	10	10	9	9	8	8	7	7	7
24	5.83	13	12	11	11	9	9	8	8	8	8	7
25	6.76	14	13	12	12	10	10	9	9	9	9	8
26	8.12	15	15	13	13	12	12	11	11	10	10	10
Taps		2	2	3	3	3	3	4	4	4	4	5
Reverses		0	0	0	0	1	1	1	2	1	2	1
Distance		3	1	3	3	4	4	5	5	5	7	3

12	13	14	15	16	17	18	19	20	21	22	23	24	25	26
-0.91	0.33	0.42	0.71	1.77	1.97	2.09	2.74	2.95	4.22	4.42	4.52	4.67	4.84	5.86
-8	-9	-9	-10	-11	-11	-11	-12	-12	-13	-13	-13	-14	-14	-15
-6	-7	-8	-8	-9	-9	-9	-10	-10	-12	-12	-12	-12	-12	-13
-5	-6	-7	-7	-8	-8	-8	-9	-9	-10	-11	-11	-11	-11	-12
-4	-5	-6	-6	-7	-7	-7	-8	-8	-10	-10	-10	-10	-10	-11
-4	-4	-5	-5	-6	-7	-7	-7	-8	-9	-9	-9	-9	-10	-11
-3	-4	-4	-5	-6	-6	-6	-7	-7	-8	-8	-9	-9	-9	-10
-3	-3	-4	-4	-5	-5	-6	-6	-6	-8	-8	-8	-8	-8	-9
-2	-3	-3	-4	-5	-5	-5	-6	-6	-7	-7	-7	-8	-8	-9
-2	-2	-3	-3	-4	-4	-5	-5	-5	-7	-7	-7	-7	-7	-8
-1	-2	-2	-3	-4	-4	-4	-5	-5	-6	-6	-6	-7	-7	-8
-1	-1	-2	-2	-3	-3	-4	-4	-4	-6	-6	-6	-6	-6	-7
0	-1	-1	-2	-3	-3	-3	-4	-4	-5	-5	-5	-6	-6	-7
1	0	-1	-1	-2	-2	-3	-3	-3	-5	-5	-5	-5	-5	-6
1	0	0	-1	-2	-2	-2	-3	-3	-4	-4	-4	-5	-5	-6
2	1	0	0	-1	-1	-1	-2	-2	-4	-4	-4	-4	-4	-5
2	1	1	0	-1	-1	-1	-2	-2	-3	-3	-3	-4	-4	-5
3	2	1	1	0	0	0	-1	-1	-3	-3	-3	-3	-3	-4
3	2	2	1	0	0	0	-1	-1	-2	-2	-2	-3	-3	-4
4	3	2	2	1	1	1	0	0	-2	-2	-2	-2	-2	-3
4	3	3	2	1	1	1	0	0	-1	-1	-1	-2	-2	-3
5	4	3	3	2	2	2	1	1	-1	-1	-1	-1	-1	-2
5	4	4	3	2	2	2	1	1	0	0	0	-1	-1	-2
6	5	4	4	3	3	3	2	2	0	0	0	0	0	-1
6	6	5	4	3	3	3	2	2	1	1	1	1	0	-1
7	6	5	5	4	4	4	3	3	2	1	1	1	1	0
8	7	6	6	5	5	5	4	4	3	2	2	2	2	1
9	8	8	7	6	6	6	5	5	4	4	4	3	3	2
5	5	5	5	6	6	6	6	6	7	7	7	7	7	8
1	3	2	3	3	4	4	2	2	4	4	4	3	5	6
3	6	7	6	10	10	9	8	9	11	10	11	12	10	11

These summations, item/person values, can be expressed as counts converted to proportions, that is, ratio of count correct to the total (p-values). The value for each cell in figure 1 is determined by the difference between measured person ability and item difficulty for each of the rows and columns. These values are for the KCTR.

Each cell in a persons by items matrix begins as a count, transformed to a proportion, a difference, and then to a probability of success for any person taking any item.

Figure B17.2 illustrates the above sequence of steps in a new example. The two left columns give the score groups by counts and associated logistic measure using the formula, $P = 1 / (1 + \exp(-(B-D))$ (or

$P = 1 / (1 + \exp(D-B)))$ with the items identified along the top row by their associated logistic difficulty. The matrix of cells formed by the intersection of row measure and item difficulty are reported as the probability of each calibrated measure value associated with answering correctly any calibrated spelling item in the test. Below the table matrix are the composite elements of the KCTR task, which in this example defines the structure of the KCTR by its item specification, which can be explained in narrative form as substantive theory and mathematically by its specification formula embodying the three characteristics of each KCTR item.

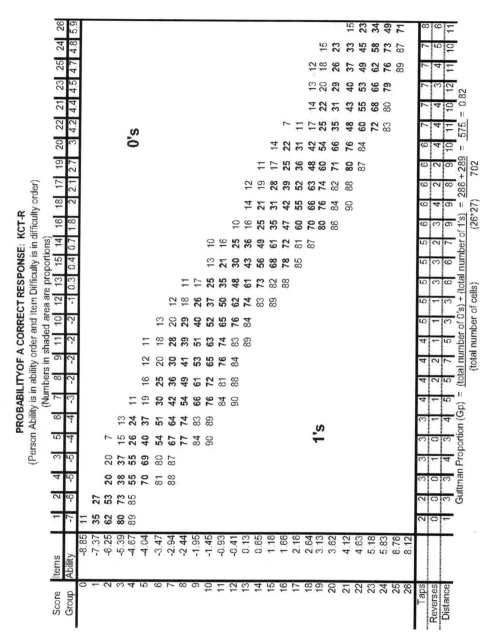

PROBABILITY OF A CORRECT RESPONSE: KCT-R

(Person Ability is in ability order and Item Difficulty is in difficulty order)

(Numbers in shaded area are proportions)

Following the strategy for the KCTR, with the spelling test, we should be able to build a specification equation based upon such characteristics of spelling words such as (1) word length, (2) regular versus irregular word construction, (3) reference to objects versus abstract qualities, parts of speech, word frequency in language, and so on. This strategy informs us that the measurement task is not complete solely by fitting data to a Rasch model. In fact, it could be said that the task onky begins as we seek to understand the item characteristics that cause the item calibrations. The task requires us to provide a narrative explaining what constitutes the substantive theory underlying the variable, and then producing a specification equation that provides a workable equation to predict outcomes based upon the characteristics isolated by the specification equation. For the KCTR, these characteristics were determined to be tapping series (length), reverses in tapping, and overall length of the tapping series. Figure B17.3 maps the item difficulties for the KCTR.

Item Difficulty

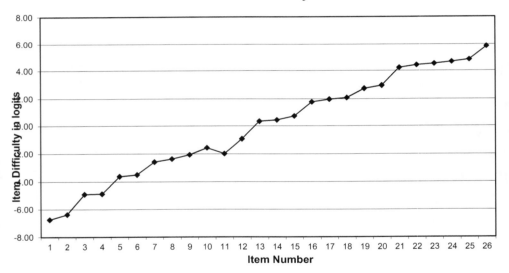

For spelling, we need to imagine (hypothesize) those characteristics that cause the spelling words to order as they do and investigate these characteristics using a causal model that specifies prediction of an outcome by the identified characteristics of spelling word difficulty - in a manner precisely like the one employed in specifying the KCTR.

It is the substantive theory and a specification equation(s) that reflects our understanding of a variable. This understanding increases with increasingly successful prediction and extension go novel item types. Figure B17.4 shows the high relationship (correlation/prediction) between the KCTR substantive theory and specification equation to the observed data. But item calibration is not the end of the process. We must go beyond calibration to examine why the items are ordered as they are. In the case of the KCTR, this has already been provided by the test characteristics given below the margin calibration values.

Stenner and Smith (1982) found a high correlation between any two of these facets.

	Taps	Reverses	Distance
Difficulty	.94	.87	.95
Taps	.82	.90	
Reverses	.90		

KCTR (Stone 2002) development produced slightly higher correlations but a similar order:

	Taps	Reverses	Distance
Difficulty	.98	.91	.94
Taps	.89	.91	
Reverses	.98		

Figure B17.4 shows how closely these theory-based item characteristics relate to observed ones. We can now proceed from theory to fill gaps in the item sequence, produce alternative ones (or eliminate duplicates), and build strictly parallel tests that all share the same correspondence table linking counts correct to measures. The success of this step allows us to move beyond mere calibration to an understanding and control of the variable of interest.

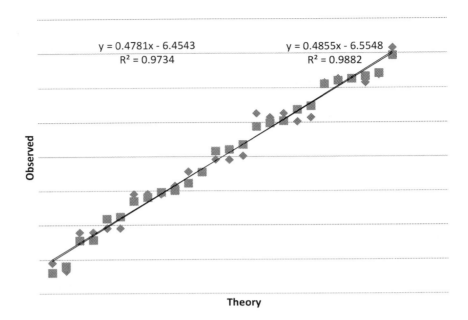

$$y = 0.4781x - 6.4543$$
$$R^2 = 0.9734$$

$$y = 0.4855x - 6.5548$$
$$R^2 = 0.9882$$

Observed

Theory

The fundamental structure of a variable can be considered the *hintergrund* or urgrund, that which lies beneath the observed scores. This process has been utilized to discern the essential characteristics in great works of art and in scientific conceptualizations. Examples can be illustrated by the analysis of musical compositions proposed by Heinrich Schenker, analysis of fictional epics by Homer and Dante, the elements of plane geometry by Euclid, taxonomy by Linnaeus, the periodic table of elements, and conceptualization of the nucleus of an atom by rings not visible. We do not focus solely upon the object of interest but upon its deeper analytic structure that gives rise to conceptualizations and investigations that go beyond the initial object of focus. Only with such procedures can we build viable variables that spring from substantive theory and specification equations. We also must focus on the continuous need to validate variables because nothing can be assumed final in such investigations. Validation is a never-ending process.

Box 18: Specific Objectivity: Local and General

Georg Rasch used the term "specific objectivity" to describe that case essential to measurement in which "comparisons between individuals become independent of which particular instruments–tests or items or other stimuli–have been used. Symmetrically, it ought to be possible to compare stimuli belonging to the same class–measuring the same thing–independent of which particular individuals, within a class considered, were instrumental for comparison" ([1960] 1980).

"Local objectivity" is the term used to designate the case in which relative measures are empirically discovered to be independent of which instrument is used to take the measures. Local objectivity is a consequence of a set of data fitting the Rasch model. When data fit is observed, differences among person measures and among item calibrations are independent of one another, and hence of the sampling of items and persons. Fit means that any two items can be shown always to differ by a statistically equivalent amount, no matter which sample of persons actually respond to the items. Similarly, any two persons can be shown to differ by a statistically equivalent amount, no matter which sample of items is used to implement the measurement procedure. Consequently, when data fit a Rasch model, the relative locations of persons and items on the underlying continuum for a construct are independent of their sampling. Absolute measures can only be obtained indirectly by introducing some reference persons or reference items of specified absolute measure into the analysis.

Since local objectivity is empirically based, it can only be statistically confirmed by further sampling. Sampled results may imply a promising variable—promising because it spaces items into a useful substantive hierarchy of "meaningful moreness." Results may encourage, even confirm, a strong intention of how items "ought" to order. But the numerical specifics of the item difficulties are estimates from this sample of persons. They may turn out, on further sampling, to be statistically reproducible, but their quantities cannot be deduced other than by reference to empirical data.

An ideal, approximated by instruments in physics and chemistry, is that absolute calibration of an instrument, such as a thermometer, does not require the services of an object or another instrument. The location of an object on, say, the Celsius scale, is not instrument dependent (i.e., any "thermometer" will do). Neither does that location depend on measuring any other object. Temperature theory is well enough developed that thermometers can be constructed and calibrated without reference to any object, or any other thermometer. Measurement of the temperature of two objects results in not just instrument independence for the difference between their temperatures but also instrument independence for the amount of each object's temperature measure.

Absolute measures are obtained directly when a specification equation, which implements a theory, calibrates instruments with useful and reproducible precision. The instrument calibrations are then based not on the measurement of any objects but on the design and efficiency of the specification equation and the quality of the manufacturing process.

The term "general objectivity" is reserved for the case in which absolute measures (i.e., amounts) are independent of which instrument (within a class considered) is employed, and no other object is required. By "absolute," we mean the measure "is not dependent on, or without reference to, anything else; not relative" (Webster 1972).

The table below contrasts local and general objectivity. "The differences between *local* and *general* objectivity are seen not to be a consequence of the fundamental natures of the social and physical sciences, nor to be a necessary outcome of the method of making observations, but to be entirely a matter of the level of theory underlying the construction of the particular measurement instruments" (Stenner 1990).

Anatomy of Objectivity		
Aspect	**Local objectivity**	**General objectivity**
Basis	empirical	theoretical
Data	sampled, exploratory, effects unknown	constructed, specified, effects known
Philosophy		
Intention	exploration	measurement
Construct Definition	incomplete, data-discovered	complete, theory-specified
Observations	quantified by data-based inference	quantified by theory-based specification
Origin, Unit, Precision	sample-estimated, varies	theory-specified, fixed
Item Calibration		

Calibrations	sample-estimated differences	theory-specified amounts
Misfit Diagnosis	depends on person and item sampling	construct consistency – independent of person and item sampling
Person Measurement		
Measures	sample-estimated differences	theory-calculated amounts
Misfit Diagnosis	item-by-person confounded	person-specific
Meaning		
Criterion	implied by sampled items	defined by theory
Norms	sample and test specific	general to the scale
Meta-analyses	aggregates indices of relative effects	aggregates amounts in measured units
Manufacturing		
Test Costs	person/item sampling, many iterations, expert supervision and evaluation	person and item targeting, routine review
Equating and Banking	sampled from common persons or items	prespecified by common theory
Quality Control	expert evaluation, person-by-item confounding	predesigned routine

Specific objectivity—local and general. Stenner, A. J. 1994. ... *Rasch Measurement Transactions* 8 (3): 374.

Box 19: Quality Control, Popular Measurements

Quality assurance in testing has been approached in two very general ways. The first has been to assure that test materials are only available to those with the appropriate level of education and training. Most publishers of assessment tools and tests use criteria to assure that only appropriately trained persons have access to their instruments. The second has been to recommend test development procedures following the Standards for Educational and Psychological Testing (AERA 1999). Both approaches are necessary but not sufficient to assure that quality can be maintained in testing. Neither one provides assurance that tests of variables are adequately built and maintained. Neither method meets quality assurance standards. To do so requires specific attention to quality control (Stone 2000). Interestingly, the index to the Standards does not include quality or quality control as headings, but statistical quality control has long been employed to assure the highest standards in manufacturing goods.

The earliest and most systematic exposition of quality control was given by Walter Shewhart of Bell Laboratories (1931, 1986). His efforts have been propagated through the lectures and writings of W. Edwards Deming, well known for his own work in quality control. The problem of quality control in testing has been frustrated by several fundamental conceptual issues. The first is addressed by Deming in his introduction to the reprint of Shewhart's Statistical Method from the Viewpoint of Quality Control (1986). Deming writes:

> There is no true value of anything. There is, instead, a figure that is produced by application of a master or ideal method of counting or measurement. This figure may be accepted as a standard until the method of measurement is supplanted by experts in the subject matter with some other method and some other figure. (ii)

Deming goes on to point out that all values and constants are in error because they are conditioned by the methods of their determination. "Every observation, numerical or otherwise, is subject to variation" (1986, ii). However, there is useful information in variation. The issue is not just error, but control over error. The second issue raised is the need of a method for establishing quality control. Error must be brought under control if the resulting values are to have any practical use. Shewhart's model for statistical control over error requires answers to these five questions:

1.   How are the observations to be made?
2.   How are the samples to be drawn?
3.   What is the criterion for control?
4.   What action will be taken as a consequence?
5.   What quantity of data is required?

He arranged these questions into a dynamic model (see next page):
Specification -----> Production ------> Inspection
(Their arrangement is best seen as a circle—or even better as a continuing spiral.)

Accuracy and Precision

Quality control in testing requires addressing accuracy and precision. The concepts of accuracy and precision in testing and measurement are called validity and reliability. The first and most important matter is to determine what any concept means. Bridgman (1928) specified what has come to be known as an "operational definition," which serves as the mechanism for understanding what a concept means. He indicated that "The concept is synonymous with the corresponding set of operations" (5). Concepts are defined, explicitly or implicitly, by a methodology. A concept equals the method that describes it and vice versa. Concepts without methods are nonsense and the bantering about of concepts without considering methods is irresponsible and unscientific. Therefore, concepts such as accuracy and precision or validity and reliability cannot be separated from the methods of their determination. To be specific, we cannot speak of validity or reliability but only of some method of determining validity or reliability specified on some occasion.

A specific example of the confusion over concepts can be observed when reading reports of a test's "validity." The determination of validity is situational and not extensive. The validity of the test is conditioned by a point in time and the setting in which it took place (i.e., the methodology and sample producing the value(s)). It is an inferential leap to assume that what occurred in one circumstance has any application to another circumstance, and it is even less likely to expect it to apply to every other instance or to all other circumstances. Concepts such as validity and reliability require more careful inquiry. Specifically, a determination of validity or reliability needs to be operationally decomposed into two important facets: the contribution from the items and contribution from the persons. Typically, and all too often, studies of test validity and reliability fail to provide any coefficient resulting from use of the sample.

For determining reliability, the KR20 is often calculated for items but almost never for persons. Hoyt (1941) recognized both approaches, saying that "extended examination of the 'among items' variance would make it possible to decide on the heterogeneity of the respective difficulties of the items while a more extended examination of the 'among students' variance would make it possible to answer certain pertinent questions regarding the individual differences among students" (41). His good advice is almost never followed. Jackson (1939), Hoyt (1941), Alexander (1947), and Guilford (1954) have all proposed an analysis-of-variance approach to estimate reliability. The advantage of this strategy is that "test reliability" can be decomposed into the variance due to examinees, the variance due to items, and the remainder or error variance. This more complete analysis is in keeping with a quality control process in testing.

Wright and Stone (1999) have demonstrated that these matters can be even better resolved using Rasch measurement techniques, which are explained in Best Test Design (Wright and Stone 1979, 151–166); all the analyses discussed below can be produced using WINSTEPS (Linacre 2000). The shortcomings of using raw scores are remedied when a Rasch measurement analysis is made of the same data and reliability is calculated from Rasch values. In addition, Rasch measurement provides the standard errors for each person and each item. These individual errors can be squared and summed to produce a correct average error variance for the sample or any subset of persons and for the items or any subset of items. When these results are substituted for those in the traditional KR20 formula, the result is a new formula, equivalent in interpretation, but giving a better estimate of reliability than any other value produced by using raw scores. Deming's adage of progressive improvement by better methods in quality control is clearly demonstrated through applying these methods.

Shewhart (1986) also spoke of prediction as an important aspect of quality control. "Every meaningful interpretation involves a prediction" (92) and "Knowledge in this sense is a process or a method of predicting an ideal" (104). The element of prediction makes scientific results useful. In the application of a test, it is the characteristics of the new sample to which we intend to apply the test, rather than simply the description of a previous sample, that is our focus. We want to know how the test will work with the new persons who are about to take it, not the earlier iterations. We want a reliability coefficient that applies to the people we intend to test, not one that only describes the people who were previously tested. But we can actually predict the reliability for a new sample if we postulate the mean and variance for that sample. One can use these statistics and the Rasch targeting formula to calculate the reliability of the test in its new application (see Wright and Stone 1979, 129–140).

Deming, as quoted above, indicated that new methods can supplant old ones when they provide better methods and values. The Rasch separation index is one such method for producing a more useful value. Correlation-based reliability coefficients are nonlinear. The increase in reliability from .5 to .6 is not twice the improvement in reliability from .9 to .95. In fact, the increase from .9 to .95 is actually about twice the improvement in precision of the previous figures. The Rasch Separation Index (G) is the ratio of the unbiased estimate of the sample standard deviation to the root mean square measurement error of the sample. It is on a ratio scale in the metric of the root mean square measurement error of the test for the sample postulated. The Separation Index quantifies "reliability" in a more direct way with a clear interpretation.

Separation G = SDT/SET
SDT = The expected SD of the target sample
SET = The test standard error of measurement for such a sample, which is almost always well approximated by SET = 2.5 / ~L

SET can also be estimated as SET = ~(C/L) where L is the number of items in the test and C is a targeting coefficient (see Wright and Stone 1979, 135–136). A figure given below expedites applying this procedure (see pp. 22–23 for remaining figures).

The Standards (AERA 1999) in section 13.14 recommend, "score reports should be accompanied by a clear statement of the degree of measurement error associated with each score" (149). Rasch measurement analysis routinely provides standard errors for every possible test measure along the variable that fully meets this recommendation. If reliability, as defined by the Standards, is the degree to which test scores are free from errors of measurement, then it follows that every ability measure should be accompanied by a standard error as an index of the degree to which this criterion is met for that measure.

The Rasch measurement standard errors satisfy this recommendation by providing individual errors of measurement for every observable measure. Where a collective index of reliability is desired, the Rasch Separation

Index is even more useful than the traditional indices of reliability. Figure 2 describes the Rasch analysis of a response matrix and figure B19.3 describes the computation of the Rasch person separation index. The targeting coefficient C varies between four and nine depending on the range of item difficulties in the intended test and the target sample's expected average percent correct on that test. Figure 4 gives some values of C for typical item difficulty ranges and typical target sample mean percent correct. However, it is not the algebraic and statistical similarity of the KR20 and the Separation Index C that is of major importance. Instead it is the decomposition of these single indices into their constituent parts leading to a more detailed and more useful management of information. Quality control is now operating.

With Rasch measurement analysis, we are able to obtain the standard error of calibration for each individual item as well as the standard error of measurement for each person's ability. With traditional methods, a single standard error of measurement is provided and only for measures at the group mean of person ability. The standard error specific to each item or person statistic is far more useful than any single sample or test average.

The location of each item and person on a line representing the variable together with standard errors provides definition and utility to the test variable. The definition of the variable is specified by the location of the items. The utility of a test variable for measuring persons is quantified by the standard error that accompanies each individual person's measure.

A variable can be thought of as a straight line. To measure successfully, we must be able to locate both items and persons along this line. A simple example is where items are located by the number of persons getting specific items correct. Persons are located by how many items they were able to answer correctly. Items to the left side of the line are easier than those to the right, while persons to the left have less ability than others to the right.

It is necessary to locate persons and items along the line of the test variable with sufficient precision to "see" between them. Items and persons must be separated along this line for useful measurement to be possible. Separation that is too wide usually signifies gaps among item difficulties or person abilities. Separation that is too narrow, however, signifies redundancy for test items and not enough differentiation among person abilities to distinguish between them. Items must be sufficiently well separated in difficulty to identify the direction and meaning of the test variable. To be useful, a selection of items (i.e., a test) must separate relevant persons by their performance. The item locations are the operational definition of the variable of interest, while the person locations are the application of the variable to measurement. Such an approach meets Bridgman's requirements for an operational definition.

Conclusion

Item and person separation statistics in Rasch measurement provide analytic and quality control tools by which to evaluate the successful development of a variable and by which to monitor its continuing utility. Successful item calibration and person measurement produces a map of the test variable (Stone, Wright, and Stenner 1999). The resulting map is no less a ruler than the ones constructed to measure length. The map indicates the extent of content, criterion, and construct validity for the test variable. Empirical calibration of items and measures of persons should correspond to the original intent of item and person placement. Changes must be made when correspondence is not achieved. Rasch measurement provides the quality control tools needed in testing.

There should be continuous dialogue between the plan for the test, the item calibrations, and person measures. Test variables are never created once and for all. Continuous quality control is required in order to keep the map coherent and up to date. Support for reliability and validity does not rest in coefficients but in substantiating demonstration of relevance and stable indices for items and measures. Such procedures assure quality control in maintaining the test variable and assuring its relevance.

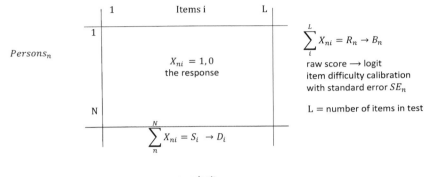

$$\sum_{i}^{L} X_{ni} = R_n \rightarrow B_n$$

raw score ⟶ logit
item difficulty calibration
with standard error SE_n

L = number of items in test

$$\sum_{n}^{N} X_{ni} = S_i \rightarrow D_i$$

raw score ⟶ logit
item difficulty calibration
with standard error SE_i

N = number of persons in sample

Items: Logit item difficulty calibration:

$$D_i = \sum_{n}^{N} B_n / N_n + [1 + (SDB^2 / 2.9)]^{\frac{1}{2}} \log[(N - S_i) / S_i$$

$$SDB^2 = \sum_{n}^{N} (B_n^2 / N) - \left(\sum_{n}^{N} B_n / N\right)^2$$

Logit item difficulty calibration error variance:

$$SE_i^2 = [1 + (SDB^2/2.9)].[N/S_i \ (N - S_i)]$$

Persons: Logit person ability measure:

$$B_n = \sum_{i}^{L} D_i / L_i + [1 + (SDD^2 / 2.9)]^{1/2} \log[R_n / (L - R_n)]$$

$$SDD^2 = \sum_{i}^{L} (D_i^2 / L) - \left(\sum_{i}^{L} D_i / L\right)^2$$

Logit person ability measure error variance:

$$SE_n^2 = [1 + (SDD^2 / 2.9)][L / R_n(L - R_n)]$$

$2.9^{1/2} = 1.7$ = scaling factor between logistic ogive and normal ogive

Rasch Analysis of Response Data

182

(See PROX estimation formulas, Wright and Stone 1979, 21-22.)

$$G = STB/RMSEB \quad \text{where}$$

$$STB^2 = SDB^2 - MSEB$$

$$SDB^2 = \sum_{n}^{N} B_n^2 / N - \left(\sum_{n}^{N} B_n / N \right)^2$$

$$RMSEB^2 = MSEB = \sum_{n}^{N} SEB_n^2 / N$$

$$B_n = \text{logit measure of person } n$$

$$SEB_n = \text{standard error of } B_n$$

so $G^2 = R/(1 - R)$ and $R = G^2/(1 + G^2)$

and $R = 1 - (MSEB/SDB^2)$ is

$$\approx 1 - (VR/VS) = [(L - 1)/L]KR20$$

with VR and VS as defined in Figure 19.1

note: $MSEB = C/L$ in which $4 < C < 9$
and $C = 5$ or 6 is typical.

Rasch Person Separation Index

(See Wright and Stone 1979, 134-136.)

Test Item Difficulty Range in Logits

Expected Percent Correct of Test Sample

	1	2	3	4	5	6
50	4.0	4.4	4.8	5.3	5.8	6.8
60	4.4	4.4	4.8	5.3	6.2	6.8
70	4.8	5.3	5.3	5.8	6.8	7.3
80	6.2	6.8	6.8	7.3	7.8	8.4

$$SET = \sqrt{\frac{C}{L}}$$

L = Number of Items in Test

(See Wright and Stone, 1979, p. 214)

If an expected reliability is also desired, it can be obtained from: $R = G^2/(1 + G^2)$.

Rasch Separation Indexes	Corresponding Reliability Coefficients
$G = \sqrt{[R / (1 - R)]}$	$R = G^2 / (1 + G^2)$
1	0.50
2	0.80
3	0.90
4	0.94
5	0.96

Values of the Targeting Coefficient C

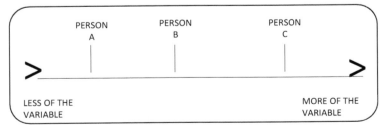

Once the variable is constructed by the line/hierarchy of items, we can proceed to locating students on this same line. Their probable positions can be specified initially by our best guess as to their ability to correctly answer the items that define the variable. The line of our variable shows both the positions of items and the positions of students. Eventually the positions of students will become more explicit and more empirical as we observe what items they correctly answer.

Consider this picture:

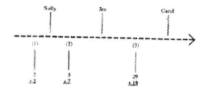

Sally's position on the variable is indicated by an expected correct response to item 1 but expected incorrect responses to items 2 and 3. Her differing responses to items 1 and 2 locate her on the variable between two items that describe her ability in arithmetic computations. She can add 2 and 2 but not 5 and 7.

Jim's position is between items 2 and 3 because we expect him to answer items 1 and 2 correctly but not item 3. In Jim's case, we have somewhat less precision in determining his arithmetic ability because of the lack of items between items 2 and 3. If we had additional items in this region, we could obtain a more accurate indication of Jim's position on the variable as defined by his responses to these additional items.

Box 20: Data Manufacturing RMT

1. Data are presumed valid and useful, so that subsequent analysis will be productive and results informative. While careful effort is usually given to collection of data, systematic checking for data quality is often omitted. When quality control is overlooked, results may be suspect.

2. Collecting data asserts that what is utilized has already passed scrutiny. It suggests that data is already in a pristine, though unsubstantiated, state. Is this presuming too much? Does *collecting* data provide what is required or does it open the door to contamination? The process of "collecting" is questionable as a basis for assuring quality data.

3. Manufacturing data describes a more useful kind of process for our purposes. Rather than "find" useful data existing in nature, they are manufactured for the occasion. Data cannot be found laying around in their natural state, waiting to be picked up by those who stumble

upon them. Data are fictitious, standing for instances of information in the solution to a problem, produced from a sample, and inferred to a population. While everyone realizes we cannot know the parameters, few would want to doubt the statistics computed from their own data. Yet we should be doubtful, and the only way to assuage that doubt is to scrupulously address the manufacturing task.

6. The production of data requires scrutiny. We have an idea and then an intention. We plan and carry out a strategy. We choose the specifications and manufacture according to our intentions. The process of data manufacture should be no less rigorous than subsequent data analysis. Quality control over production is absolutely necessary. We should keep records on data manufacturing and monitor the process with control charts. Rasch measurement strategies have always stressed data scrutiny. The examination of item and person records, fit analyses, and data plots are some of the tools by which Rasch measurement monitors the quality of manufactured data.

Box 21: Item Specification versus Item Banking

Our thesis is simple and straightforward. It is not necessary to have a bank of items for measuring a construct when we possess an algorithm for writing an item at any desired level of difficulty. The algorithm is the *key to the bank,* so to speak. If one has the key, the bank is open.

Bruce Choppin (1968) was an early Rasch pioneer who promoted item bank development. Items representative of the variable of interest are banked and selected for use as required. Leveled paper-pencil tests can be quickly assembled from the bank of items based on their associated item calibrations and item use histories. Also, computer-based adaptive tests can be assembled electronically and targeted to each examinee. As useful as item banking has proven to be, it is possible to move beyond the banking of individual items and their associated item statistics.

When enough is known about what causes item difficulty, a specification equation can be written that yields a theory-based item calibration for any item the computer software designs. An item's calibration is seen to be the consequence of decisions the computer software makes in constructing the item. This process mimics the steps a human item writer takes in constructing an item, albeit with more control over the causal recipe for item difficulty. A thesis of this paper is that when asserting that a measure possesses construct validity, there is no better evidence than demonstrated experimental control over the causes of item difficulty.

A measurement instrument embodies a construct theory, a story about what it means to move up and down a scale (Stenner, Smith, and Burdick 1983). Such a theory should be vigorously tested. In a demonstration of these methods, Stone (2002c) theorized that the difficulty of short-term memory and attention items (Knox Cube Test) was caused by (1) number of taps, (2) number of reverses in the direction of the tapping pattern, and (3) total distance in taps for the pattern. This theory was tested by regressing the observed item difficulties on the above-mentioned three variables. The figure plots the correspondence between predicted (theoretical) item difficulties and observed item difficulties. Ninety-eight percent of the variation in observed item difficulties was explained by number of taps (standardized Beta = .80) and distance covered (standardized Beta = .20). Number of reverses in the context of these two predictors made no independent contributions. An earlier study (Stenner and Smith 1982) using different samples of items and persons found that an equation employing the same two variables explained 93 percent of the item difficulty variance. Finally, Stone (2002c) reanalyzed KCT-like items developed over the last century and found a striking correspondence between the two variable theory and observation. We should note that there is some uncertainty in the observed item difficulties analyzed in these studies, suggesting that the dis-attenuated correlation between theory and observation approaches unity.

When item difficulties and, by implication, person measures are under control of a construct theory and associated specification equation, it becomes possible to engineer items on demand. No need to develop more items than you need, pilot test these items, estimate item calibrations, and then bank the best of these items for use on future instruments. Rather, when an instrument is needed, an algorithm generates items to a target test specification along with calibrations for each item, and a correspondence table links count correct to a measure.

Applications that incorporate the above ideas are under development for the next KCT revision and for an online reading program that builds reading items real time as the reader progresses through an electronic text.

Some of the practical benefits of what might be called theory-referenced measurement are: (1) if the process yields reproducible person measures, then evidence for construct validity is strong; (2) test security is facilitated because there are no extant instruments that would be compromised upon release; and (3) a fully computerized procedure keeps the process under tight quality control at a fraction of the cost of traditional item standardization procedures.

Finally, one well-recognized means of supporting an inference about what causes item difficulty is to experimentally manipulate the variables in the specification equation and observe whether the predicted item difficulties materialize when examinees take the items. In building the latest version of the KCT a part of the scale had an insufficient number of items. The specification equation was used to engineer candidate items to fill in the space. Subsequent data collection confirmed that the items behaved in accord with theoretical predictions (Stone 2002b). Although this exercise involved only four KCT items, it suggests that the construct specification equation is a causal representation (rather than merely descriptive) of the construct variance.

Reflecting on this extraordinary agreement between observation and theory suggests two conclusions: (1) the specification equation affords a nearly complete account of what makes items difficult, and (2) the Rasch model used to linearize the ratios of counts correct/counts incorrect must be producing an equal interval scale or a linear equation could not account for such a high proportion of the reliable variation in item difficulties.

Measurement of constructs evolves along a predictable course. Early in a construct's history, measurements are subjective, awkward to implement, inaccurate, and poorly understood. The king's foot as a measure of length is an illustration. With time, standards are introduced, common metrics are imposed, artifacts are adopted (e.g., the meter bar), precision is increased, and use becomes ubiquitous. Finally, the process of abstraction leaps forward again, and the concrete artifact-based framework is left behind in favor of a theoretical process for defining and maintaining a unit of length (oscillations of a cesium atom). Human science instrumentation similarly evolves along this pathway of increasing abstraction. In the early stages, a construct and unit of measurement are inseparable from a single instrument. In time, multiple instruments come to share a common metric, item banking becomes commonplace, and finally, the construct is specified. When a specification equation exists for a construct and accounts for a high percentage of the reliable variance in item difficulties (or ensembles), the construct is no longer operationalized by a bank of items but rather by the causal recipe for generating items with prespecified attributes.

SUBSTANTIVE SCALE CONSTRUCTION

7

Constructing "yardsticks" to make measures requires a close collaboration between statistical techniques and substantive hypotheses. The partnership between item/person data and the intention of the scale is important. Critics of item writing have lamented the lack of models for the construction of test items (Guttman 1969; Cronbach 1970). Rasch measurement addresses the relation between the substance and the process of item construction in building a yardstick. Yardstick construction begins by: (1) establishing a design for covering the sequence of content to be sampled by the items, (2) writing the response directions and items implied by this design with clarity.

A test is an aggregate of items. But the items must be ordered from easy to hard to guide construction and analysis. Achievement tests follow a specific plan designed to cover the range of designated subject matter. Questionnaires and rating scales are designed to assess experiences, feelings, attitudes, and opinions. The task of building an instrument begins with constructing items to correspond to an ordered definition of the intended variable (Stone 2002a). Since the idea of a variable is sequential, the items should parallel this sequence. For example, to measure arithmetic computation, each proposed item must be harder, easier, or of the same difficulty as another along a single line of inquiry. Items must be constructed by their difficulty and correspond to the expectations for person ability. More able persons are expected to respond successfully to difficult items than less able persons.

For attitude scales, persons with "more" of the attribute are expected to respond more readily to items demanding "more" of the attribute (its difficulty). Consequently, items responded to more commonly are located at the "easy" end of the scale, and items less frequent or rare in response are at the "difficult" end. It is important to be clear about what we want to determine about an attitude and choose words that clearly indicate what is expected.

So-called bipolar response alternatives appear useful but are not so because they combine two scales into one. Agree or Disagree (in some amount) are not equal opposites. Each direction using a bipolar response alternative may imply a different mind-set and constitute a separate scale. Movement away from the center depends upon a respondent's attitude about the statement. It cannot be assumed that respondents move in opposite directions in equal units simply because the response alternatives progress numerically or by using the words "somewhat" to "a lot." Bipolar response alternatives only introduce confusion into a scale. These problems are rarely recognized and analyzed.

Linking

A set of items, a test, a questionnaire or a rating scale, is constructed by sequencing individual items along a single underlying dimension. Sequencing is satisfied when the set of items operate as a cohesive whole. Then both items and persons can be located on one "line" in one direction of "more." The essential characteristic of this line is the conjoint transitivity of item and person parameters.

If we have two tests of different items, they can be "linked" by the addition of a shared set of common items or by a shared sample of common persons. The linking value from common items is the difference between the average within-test-calibrations of the linking items in each test. A linking value increases in reproducibility as the number of items in the "link" is increased.

If two tests, A and B, are joined by a common link of K dichotomous items and each test is given to its own sample of N persons, then item calibrations d_{iA} and d_{iB} can represent the estimated difficulties of item i in each test with standard errors of approximately $2.5 / N^{1/2}$. The single constant necessary to translate all item difficulties in the calibration of Test B onto the scale of Test A can be estimated

$$G_{AB} = \Sigma_i(d_{iA} - d_{iB}) / K \ i = 1, K$$

with a standard error of approximately $3.5 / (NK)^{1/2}$ logits.

The quality of this link can be evaluated by the fit statistic:

$$\Sigma_i(d_{iA} - d_{iB} - G_{AB})^2 \left(N^{1/2}\right)[K/(K-1)] \sim X_K^2 \ i = 1, K$$

which, when two tests fit together, will be distributed approximately chi-square with K degrees of freedom.

The fit of an individual item can also be evaluated,

$$(d_{iA} - d_{iB} - G_{AB})^2 (N^{1/2}) [K/(k-1)] \sim X_1^2$$

When the performance of that item is consistent with the equating, it too will be approximately chi-square distributed with one degree of freedom. Karabatsos (2000) provides a useful discussion of fit statistics, the detection of measurement disturbances and the properties of an accurate fit statistic.

Building Yardsticks

The task of building an instrument, a yardstick, requires calibrating each item on a common line which defines its location relative to all other items on this line as more difficult, less difficult; or more common, less common. The span of items should extend beyond the range of the expected respondents' measures.

Substantive scale construction requires hypotheses about a conjoint item and person order to be defined by the yardstick. While item construction is usually considered paramount in scale construction, it is also necessary to consider the persons to be measured. We need to specify the type(s) of persons the yardstick is intended to measure and the range of measures expected. All relevant person characteristics (age, gender, etc.) must be identified, and a plan made for labeling persons accordingly. In order to settle questions of bias and differential item functioning (DIF), person labels must identify every person characteristic we hypothesize to matter (e.g. gender, age etc.). We must also consider the format in which items are presented and the way in which we expect persons to respond. The response format needs to fit the persons of interest.

Spelling

Suppose our task is to measure the spelling ability of students in grades one to three. While we have a grade/age range in mind, we must also take into account that some students will exceed this range because they have more spelling ability than our target group. Other students might have less spelling ability. Consequently, our item difficulties need to encompass persons above and below our target group in order to measure everyone that we expect to encounter.

Next, we decide how the "spelling" items will be chosen. Select a spelling word expected to be of middle difficulty for the target group of students. Such a word can be obtained from teachers of these grade levels or by consulting an age/grade level and word difficulty compendium. Suppose "meeting" is chosen as the first word. This word is placed in the center of what will become a hierarchical list of spelling words. Teachers are then asked to select a harder word, followed by an easier word. Suppose we have dog, response, and meeting

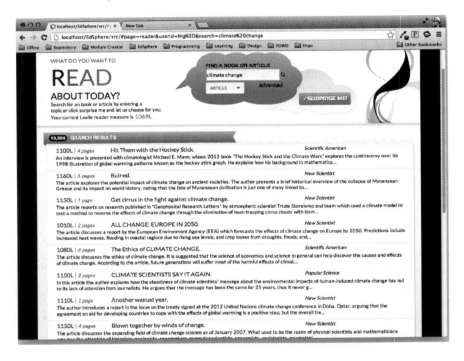

as the first three words on our spelling yardstick. It becomes straightforward to add new words, harder than "response"; easier than "dog" and in between "dog" and "meeting"; "meeting" and "response." The exercise continues as additional words are added to the right and left of each word chosen. Further validation can be obtained from other teachers, from curriculum experts and word frequency compendia. Some words may be thought to be at the same level of difficulty. These can be considered redundant and eliminated, or saved and made available to alternate forms of the instrument. Some words might be retained because of their substantive importance, irrespective of their expected difficulty, but their retention should be monitored.

The end result is a continuum of spelling words arranged from easy to hard. They encompass the range of the target group but also include more difficult words for more able persons and easier words for those with limited spelling ability.

Next, determine the mode of administration. A spelling test can be administered by oral dictation with the teacher reciting the word to be spelled, by multiple choice in which the respondent chooses the correctly spelled word from among four or five choices, or a cloze method, by which the correctly spelled word must be discovered (a vocabulary challenge) and then spelled correctly in the space provided in order to complete a sentence. The response format must be carefully planned so that there is no contamination of administration mode with item difficulty that cannot be controlled. The research design must consider the method of administration, response mode, item and person characteristics. Monitor the subsequent item calibrations and person measures against modes of administration and other characteristics deemed important in order to determine whether the outcome corresponds to your expectations.

Depression. Suppose we seek to measure the amount of depression people self-report. The target range encompasses those individuals with little or no measurable depression to those with deep and explicit depression. A team of depression experts is consulted to construct a range of relevant indicators, item "germs," to index varying amounts of depression. An item "germ" is the initial idea for an item in capsule form (Stone 1997). Acquaintance with the literature on depression will help identify item germs that cover the topic. This will verify and improve upon the opinions of the experts and contribute to the validity of the items. At this point, however, postpone the exact composition of the items and simply use item germs as "indicators" to complete the hierarchical specification of the depression yardstick.

The strategy is similar to the one used for spelling. Begin with an indicator expected to be in the middle of the scale, say, "remorse." From this indicator, add another indicator suggesting more depression such as "suicide," and another suggesting less depression such as "fitful sleeping."

We might notice that we are mixing cognitive, mood, and physical manifestations of depression. This may lead us to develop three yardsticks. If so, we will want to construct a hierarchy for each one. The items may be mixed for administration but separated for analysis and reporting. Multiple scales can help organize the indicators of whatever the essential components of depression are determined to be. Since

Rasch measurement specifies a unidimensional scale, this is sometimes misconstrued to mean that all the items on an instrument must always be of one and the same content. This is not so. What is unidimensional in Rasch measurement are the conjoint probabilities of items and persons for a specific set of items and a relevant sample of persons. It is reasonable to write items to follow a single line of inquiry according to a stated goal. But, a variety of item types can be combined in an administration when it is useful to do so. Once we have our item germs, our indicators, in hierarchical order, we can address how to write out each indicator in its best presentation. The final form of an item, however, can shift its location in the hierarchy. Consider "I have had so many periods in my life when I was so cheerful and used up so much of my energy that I fell into a low mood." The purpose of the item appears to be determining a "low mood," but only as a consequence of two quite different prior conditions, "so cheerful" and "so much energy," both of which are opposites to a low mood. Can the respondent easily determine that "low mood" is the focal point? Will the respondent become confused by the distracting phrases about energy and cheerfulness? A simpler phrase would be more useful, perhaps, "I am in a low mood." The first and longer sentence invites confusion. The latter keeps it short and simple. Item phrasing must be carefully reviewed. We can inadvertently introduce words of greater reading difficulty, increase the phrase length so that it leads to confusion, or use negative statements that change the balance of the item. Special attention should be given to negatively phrased items. It is often thought that a simple reversal of negatively phrased items is all that is required to bring them into conformity with the positive items. However, if items phrased positively can be confusing, how much more confusing might we expect negatively phrased items to be? Negatively stated items are seldom the reverse of positive ones. The only clear-cut way to resolve this problem is to evaluate statistically whether or not the negative items constitute a separate dimension.

Table 23.1 from *WINSTEPS* (Linacre 2002) plots a principal components analysis of item response residual similarities. These factor loadings need to be investigated when groups of items cluster together. Residual principal component outliers will reveal whether reversing negative items has been successful.

194

Item Try-Outs

Traditional item try-out practice recommends a small sample to check the final form. I suggest that you begin with yourself. Reading your items out loud into a tape recorder and then listen to how all the items sound to you. When you are satisfied with the way your items sound as you listen to your reading of them, give them to only one person. Ask that person to read and respond to each item out loud. Ask the person to tell you each mental step they experience while working through each item. Encourage the person to vocalize any musings while considering an option as well as any specific reasoning undertaken in arriving at an answer. Encourage ruminating out loud. Tape record what is said so that you can reconsider it later.

This simple step with one "typical person" will uncover most mistakes in composition, typing, confusing words or phrases and format. You only need one person to expose a typo, not twenty to thirty.

Put this information to work and make the necessary changes. Try it on a second individual. By now you will have corrected your major mistakes, and you should find there is less to change. If changes are required, make them and redo this step. Continue one person at a time until you observe that you are only getting personal affectations and favored opinions but not fundamental criticism of your instrument. Then you know your instrument is nearly error-free. With this approach, the traditional, costly "try-out" is unnecessary.

Arithmetic

Suppose we want to assess the four major operations with whole numbers taught in grades one to six. The range of usage could be even wider. It might need to be administered in sequential sections, by grade, so as to keep the rest from becoming too long, and with some items too hard for the early grades, and others too easy for the upper grades. The test can be divided into forms, each sequential form composed of relevant items. Figure 7.2 illustrates this procedure using an abbreviated sequence taken from the format used in KEYMATH (Connolly, Nachtman, and Prichett 1988).

This four-level format decomposes arithmetic operations with whole numbers into four components for the purpose of analysis and reporting. It also shows which items in each scale have similar difficulties. Such a format gives a

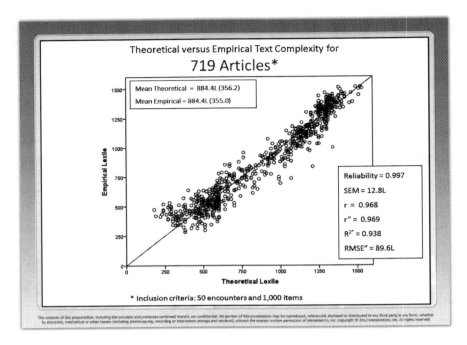

view of the entire test sequence as well as each separate scale. We can see how different operations relate to one another by difficulty. This outline also helps plan instruction.

Achievement

Another example comes from the yardsticks for the *Wide Range Achievement Test* (WRAT, Wilkinson, 1993) shown in Figure 7.2. There are three yardsticks, one measuring word naming, one measuring arithmetic computation, and one for spelling from dictation. The items from each yardstick proceed from elementary skill levels on the left to increasing grade-level difficulties on the right.

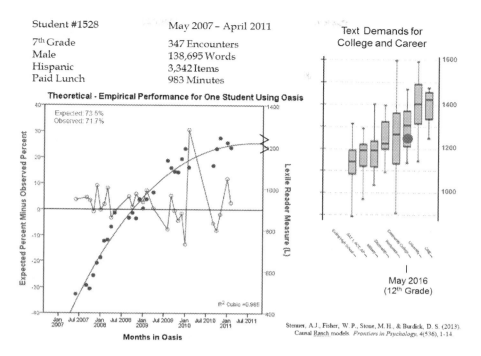

Student #1528
7th Grade
Male
Hispanic
Paid Lunch

May 2007 – April 2011
347 Encounters
138,695 Words
3,342 Items
983 Minutes

Text Demands for
College and Career

May 2016
(12th Grade)

Stenner, A.J., Fisher, W. P., Stone, M. H., & Burdick, D. S. (2013).
Causal Rasch models. *Frontiers in Psychology*, 4(536), 1-14.

The arrangement of these items is exactly what we expect to find in any measuring instrument. The initial expected locations come from teacher judgment, curriculum guides, and learning experts. The actual metric locations shown in figure 7.3 are determined by empirical evidence. Each scale goes from left to right in difficulty order. We can also make vertical comparisons between locations on any of the three scales because all three scales were equated by the same persons taking all three tests.

When the initial arrangement of the items is sensible, it will be supported by empirical evidence. When there are upsets to the initial order, we have to investigate what happened. The arrangement of persons should correspond to the arrangement of items. Persons known to be more able should be located above persons known to be less able. A hierarchical correspondence between items and persons should associate easy items with less able persons, more difficult items with more able persons. If the empirical order is confused, then it becomes clear that we do not know what we are doing.

Determining the calibrations of items and measures for persons should substantiate theory and (or) suggest revisions. A continuous dialogue must be maintained between the idea for the yardstick and experience provided by the results produced. The better the plan, the fewer the cycles required between idea and data before a usable definition of the yardstick is achieved. Successful item calibration and person measurement leads to a map of the variable. The resulting map is not less a "yardstick" than one constructed for measuring length. It provides measures as useful as those of any yardstick for measuring length.

Figure 7.3 shows the three WRAT-3 "maps." These maps are benchmarked with sample items showing the progression in item difficulty. Below the item is an equal interval scale providing measures. The location of concomitant average grade and age measures provides normative information. These maps have immediate visual appeal and application. Like the marks of increasing height of a child on the doorjamb, these maps show student progress on each of these three dimensions. Especially helpful is the overall, one-page view these maps portray. This provides a sense of order and coverage to the entire range of the three yardsticks.

Maps show the progress of pupils along the yardstick. The linear scale gives the values necessary for statistical analysis. The grade spacing mapped on each yardstick provides a normative annotation to the yardstick and also shows that growth is accelerated among younger children. The item hierarchies in each yardstick appear orderly. But the evident order is in need of continual reappraisal and cross-validation. Variable definitions are never finished. While there is consistency and order to the yardsticks comprising the WRAT3 through five editions (1947 to 1993), it remains necessary to continuously monitor these variables in order to keep these maps up to date (Stone 2002a).

Rating Scales

Attitude measures, although seemingly simple, are difficult to build. All that appears necessary is to make up some questions and add some response categories. However, writing questions, without a unifying idea and a rationale for item development and hierarchy, is not productive. Here are some guidelines for writing items for attitude scales:

1. Avoid factual statements.
2. Do not mix past and present. Present is preferred.
3. Avoid ambiguity.
4. Do not ask questions that everyone will endorse.
5. Keep wording clear and simple.
6. Keep statements short and similar in length.

7.   Express only one concept in each item.
8.   Avoid compound sentences.
9.   Assure that reading difficulty is appropriate.
10. Do not use double negatives.
11. Do not use "and" or "or" or lists of instances.

Consult Likert (1932), Payne (1951), Fink (1995), Warwick and Lininger (1975), Kerlinger (1986), and Dillman (2000) for additional suggestions.

Response Format

There are three general response formats from which to choose: dichotomies, such as "yes-no" or "correct-incorrect"; rating scales with graded responses such as "never," "sometimes," "frequently," "always"; or "none," "some," "a lot," "all"; and some form of completion such as calculating a problem, spelling a word, or filling in a blank.

The dichotomy is a simple response format and generally unambiguous to the respondent. It is easily scored. The response model is straightforward. An example is *Knox's Cube Test-Revised* (KCTR-R, Stone 2002b). Rasch's model for analysis of a dichotomy is explained in the KCT-R manual and in *Best Test Design* (Wright and Stone 1979).

Rating scales, presumed simple to construct, are actually fraught with ambiguity. For example, what distinguishes "usually" from "frequently"? Could one person's "usually" be another person's "frequently"? Do "none" and "always" mean absolutely without exception? People differ as to how they use these words. Whenever this confusion appears, it flaws the response model. But rating scales can be made to solicit information without confusion. An example is a *Liking for Science* questionnaire analyzed with a Rasch rating scale model. This is described in *Rating Scale Analysis* (Wright and Masters 1982).

Open-ended items that require respondents to write in their answer are more time-consuming to answer and to score. There is the inevitable possibility of variability in postscoring, even when very specific guidelines are established. Postscoring must be continuously monitored and reproducibility ascertained. Nevertheless, an open-ended format can be useful in determining how the task was addressed and observing the steps taken to completion. It is especially valuable to teachers evaluating the effects of instruction. The scoring model can be dichotomous (i.e., correct or incorrect) or partial credit.

Rating Scale Formats

Rating scale formats require that the category hierarchy parallel the statement hierarchy. The labeling of categories must be relevant to the item. All items may not fit one set of categories. My experience is that four options are usually enough and that two are often sufficient (Stone 1998). Pay attention to viewing the item from a respondent's point of view.

It is believed by some that having more options permits more possibilities of gradation and, consequently, increases the flow of information. But this is true only when respondents use all options of the scale in the same way. Whether this is so can be determined after the responses have been gathered. If six options are provided and respondents use only two, three, or, perhaps, four, we do not have six options to consider. What the researcher plans and what respondents do are often different. To develop a useful response format, it is important to understand and anticipate how respondents will act. The literature about a particular attitude may suggest a useful range to employ, but do your own study. Discover the useful range of responses that your respondents do employ (Linacre 2002). Do not construct a large number of questions and expect every one to fit a standard seven-point scale. Many rating scales written to require several steps actually reduce to as few as two. This will be the case when questions produce polarization. In such cases, the "average" rating computed for such an item represents a nonexistent position. If about half the respondents choose "1" and the remainder choose "5," the *average* is in the middle where no respondent indicated a preference. But if the researcher never investigates this matter, this scoring problem will go undetected.

Examples of this problem abound. Stone (1998) and Wright and Stone (2003) found this problem with respondents encountering a five-point scale when completing Wolpe's *Fear Survey Schedule.* Three response categories were most efficient, and a dichotomy was more efficient than the original five-point scale. Stone (1998) found that a dichotomy was even more efficient and more useful than the graded response scale of Beck's popular depression inventory. These examples indicate that users of rating scales must study and give more attention to the way respondents actually answer items and not base their work on unwarranted assumptions.

Fascination for the popular five-point rating scale 1-2-3-4-5 seems based more on numerology than reason. A middle response choice of "3" can reflect a decision not to prefer either end, a lack of information by which to choose, or an unwillingness to commit to a definitive response. Which is it? How can we understand what "3" means when it can indicate a variety of intentions? I see no value for a middle category given such confusion.

The use of four responses allows for a reasonable range. The responses can be analyzed as dichotomies, if desired, such as "1-2" versus "3-4"; "1" versus "2-3-4"; and so on. The middle can be eliminated leaving "1" (omit "2-3") and "4." See Stone (1998) for details.

Category labels must define a continuous increase of discriminating steps. The words chosen to articulate these steps must do so in a clear, unambiguous manner that is easily understood by all respondents. One often encounters words that appear important to the researcher but with little thought about the meaning conveyed to the respondent. But how respondents actually view the category labels is of the essence. The words that are used must have a shared, unambiguous meaning for every respondent using the rating scale.

The labels chosen should be verified. An obvious gradation must be evident to every respondent. The terminal points must be reasonable. This is another reason why the same response format cannot be universally applied to every item. Consider the question "Have you ever told a lie?" Should we expect or believe terminal responses such as "Never" or "Always"? While either response remains in the realm of possibility, is it really useful to build such response formats as this? Does "Never" indicate absolute honesty or does it provoke prevarication or flippancy. Is "Always" a believable response?

If the categories are not step-wise hierarchical, or the word labels are confusing, any resulting data must be suspect. But the literature ignores this problem. Rating scales are often constructed without considering these matters. Values are determined by multiplying an ordinal step value by the frequency count, and no further consideration is given to the problem. The resulting "scores" are treated as sufficient for statistical analysis. While these steps are often considered standard, they are inadequate. Wright and Stone (2003) give a detailed plan for how these matters can be usefully addressed.

Mapping

Common expressions such as, "See what I mean," "Seeing is believing," "Show me," "Put me in the picture," and variants testify to the primacy of vision in comprehension and communication. We need visual images to understand what we are doing and to make ourselves understood. Tufte

(1997) shows how visual images have given meaning in numerous fields of endeavor. Graphs map the data in useful ways.

Maps confirm (and correct) intentions and lay out empirical results (Stone, Wright, and Stenner 1999). Visual communication is powerful. Information gleaned from data is most effectively communicated when a picture tells the story and the data provide its support. A map delivers the idea of a yardstick by drawing out the conceptual variable to be constructed and the empirical variable that has been constructed. When yardstick construction is successful, the picture proves it. We "see" what we mean.

A hierarchy of items defines a yardstick, a visual map of a variable. But a map is not only constructed from results of a study. The conception of a yardstick is no less a map than the one constructed from data. The two maps, the idea and the thing, must collaborate.

The conceptual map of the intended yardstick is a plan for what is expected of items and persons. This map is constructed from the best information available and is based on a theory of where the items and persons are expected to be located. Theory and past experience should make it possible to indicate normative and criterion positions on the conceptual map. The conceptual map is the icon, the idea, and a hypothesis.

The empirical map results from data. If expectations are based on good theory, the subsequent results shown in the empirical map should correspond to the conceptual one. If we know what we are doing, the degree of correspondence will be substantial. If we do not understand what we are doing, the differences between the two maps will help us. Those differences will show us what to address in order to improve the correspondence between maps—between idea and experience, between icon and index. The results of the analysis displayed in the empirical map are the meaning of the data. The concept of the variable is its icon. The data are the index of the icon. The empirical map extracted from the raw data is the meaning or symbol of the icon-index relation. See Peirce (in Buchler 1940, chapter 7) for his theory of signs.

Here are some conventions for making maps.

1. An arrow points to the direction of "more" on the map.
2. Variables usually increase from left to right when pictured on the horizontal and from bottom to top when on the vertical.
3. Parallel lines (horizontal or vertical) can be used to indicate subdivisions of a yardstick such as "addition"; "subtraction," and so on.

4. A map should be pictured on a single page, and it may be modified to do so. Details can go on successive maps if necessary.
5. Items and persons (complete or representative) should be labeled and named.

   WINSTEPS (Linacre 2002) facilitates this process. The application of Rasch models via WINSTEPS (or other Rasch software) to the raw data constitutes a deduction of the empirical map from the data.

6. Both normative and criterion indicators should be located along the yardstick to add relevance.
7. The precision of measurement should be indicated on the map.
8. Misfit behavior should also be identified on the map, when relevant.

A map can be used as the basis for a self-scoring form for recording and reporting individual status. The Manuals for *Knox's Cube Test* (Stone and Wright 1980; Stone 2002b) provide examples of such a reporting form, and chapter 3 of these manuals describes its use. A misfit ruler can be made to assist this process. See chapter 8 in *Best Test Design* (Wright and Stone 1979) for details.

Data and Models

Is "collecting" the appropriate word to describe how data are accumulated? Is data collecting akin to beach-combing? Does "collecting" open the door to contamination and confusion? "Collecting data" suggests an innocence that is inappropriate for science.

"Constructing" or "manufacturing" are better terms for how data is produced. What must be clarified is the method of data production. We must build and maintain control over the production process (Stone 1996a, 1996b). Data are intended to stand for something. Data production methods are purposefully contrived in order to produce data that can represent the generalization, the idea we have in mind.

Data production requires scrutiny. We have an idea, an intention, an icon. The data must index the icon from which they were derived. We plan and carry out a strategy by choosing the specifications and fabricating the product in order to indicate and realize our idea. Data production should be no less rigorous than any other manufacturing process. Quality control over production is essential (Stone 2002a). We must keep records on the manufacturing of items and monitor the process with quality control charts. Rasch measurement strategies stress data scrutiny. Examination of item and person records, fit analyses, and data plots are the tools used to monitor the quality of data and to shape the data index so that it follows and serves the idea icon.

We do not seek models to fit data. Rather we construct data to fit the model, which is necessary in order to construct measures. Social science literature often speaks of a need to find a model to fit a particular collection of data. But the problem is, in fact, entirely opposite. The challenge is to produce data good enough to fit a measurement model. Rasch models provide the means by which to construct measures from carefully produced data and to monitor the process to reach a best possible solution.

Data production begins with observations made at a qualitative, *nominal* level. This level begins by defining each category of interest and intent, and recording which categories are nominated by which respondents. The ordering of these nominations comes from the *ordinal* positioning of the steps as designated and dictated by the researcher. The model of expected occurrence probabilities, however, is defined by *interval* measure parameters estimated by analyzing the preordered nominal data for its stochastic evidence of conjoint additivity. Measure estimation is reached by minimizing the discrepancies between parameter-based probabilities and observed occurrence frequency. Control is provided through misfit analysis that evaluates the difference between observations and expectations. The linear measures produced are accompanied by their standard errors (reliability) and their model fit (validity).

The Rasch Models

Five frequently used Rasch measurement models are the tools for constructing variables i.e., yardsticks for making measures from data. These five models have been developed for five response formats. All models share a common algebraic form. The Poisson Count, Bernoulli Trials, Rating Scale, and Partial Credit models are simple extensions of Rasch's original Dichotomous model (Rasch [1960] 1980, 1994; Wright and Masters 1982; Masters and Wright 1984). See table 1 below.

These definitions separate parameters as in:

$$G_{ni} - G_{mi} = B_n - D_i - B_m + D_i = B_n - B_m$$
$$G_{ni} - G_{nj} = B_n - D_i - B_n + D_j = D_j - D_i$$

The Dichotomous Response model records two levels of performance on an item, as in "true/false"; "correct/incorrect"; "present/absent." It is applicable to any pair of exhaustive, mutually exclusive, and orderable response alternatives. This "all or nothing" format makes it a "one-step" item. Rasch introduced this model in the 1950s, and its application can be found in Rasch ([1960] 1980, 1994).

A Rating Scale Response model in figure 7.4 supports a successive step interpretation. It can be applied to any item where ordered response alternatives are in play. The categories in such a rating scale may be printed equidistant from each other. But this hope must be verified or corrected from the behavior of respondents to the scale and not based upon the graphic imagination of the researcher.

Five Rasch Models

For n = persons,

i = items and

m = response categories:

Observe	$X_{ni} = 0, m_i$	ordinal step evidence of a variable
Estimate	$0 < P_{nix} < 1$	probability of step X_{ni}
Form $G_{nix} = \log (P_{nix} / P_{nix-1})$		log odds of the step up from $(X_{ni} - 1)$ to (X_{ni})

Define conjoint additivity:

Dichotomy	$m_i = 1$	$G_{nix} = B_n - D_i$	$\Sigma_i D_i = 0$	
Rating Scale	$m_i = m$	$G_{nix} = B_n - D_i - F_x$	$\Sigma_i D_i = 0$	$\Sigma_x F_x = 0$
Partial Credit		$G_{nix} = B_n - D_i - F_{ix}$	$\Sigma_i D_i = 0$	$\Sigma_x F_{ix} = 0$
Poisson Counts	$m_i = \infty$	$G_{nix} = B_n - D_i - \log X_{ni}$	$\Sigma_i D_i = 0$	
Bernoulli Trials	$m_i = m$	$G_{nix} = B_n - D_i - \log [X_{ni} / (m - X_{ni} + 1)]$	$\Sigma_i D_i = 0$	

Table 2

Rating Response Model

None	A Little	Quite A Bit	All	or	Strongly Agree	Agree	Disagree	Strongly Disagree
0	1	2	3		3	2	1	0

The Partial Credit model identifies one or more intermediate levels of success on particular items. Deducing the correct arithmetic operation could be level one, correctly calculating the answer level two, and properly labeling the result level three. Partial credit is assigned for reaching each of the intermediate levels. Three levels produce two steps. Four levels produce three steps, and so forth.

The Poisson Counts model counts how many times person n is successful (or unsuccessful) at item i. In principal, the count can range from zero to infinity. A person reading a passage can make as many possible errors as there are words in the passage. Rasch ([1960] 1980, 1994) used a Poisson model to analyze error counts and reading speeds on oral reading tests.

The Bernoulli Trials model addresses m independent attempts at a given target (i.e., the count x of the number of successes in m trials).

From Scores to Measures

A "measure" is an abstract number with which arithmetic can be done. Measures can be added and subtracted, with results that maintain their numerical meaning. Original observations can never be measures in this sense. All that we can actually do is to record a presence or absence of some clearly identified event. This raw count, or score, can only be interpreted as an ordinal rank. Raw observations cannot be measures because a measure implies and requires the previous construction and maintenance of an abstract quantitative system shown in practice to be useful for measuring. Since social science so often mistakes raw counts of "ordinal scores" for "linear measures" and then misuses them statistically as though they were measures, we should designate these "scores" as "counts" so that their empirical basis remains explicit, and they are not confused with "measures."

Data originate as observations. All that we observe directly is the presence or absence of a well-defined quality. What we count directly are numbers of classified occurrences. Initially all our classifications are qualitative. But some can be ordered and become more than nominal. Quantitative science begins with identifying conditions, events, and qualities deemed worth counting. These counts are the essential raw data for the construction of measures. But they are not yet measures because they do not have the numerical properties necessary to support arithmetic. Counting is the beginning of quantification. Measurement is constructed from well-defined sets of counts.

The relationship between the raw counts produced by a fixed set of observations and the measures they may imply is ogival. The necessarily finite interval between the minimum observable count of "none" and maximum observable count of "all" must be extended to imply an infinite interval of measures. Figure 3 shows this relationship between counts and measures. Near the center of the ogive, at 50 percent, the relationship between raw counts and measures (for complete data) looks approximately linear. But this relationship tends to flatten out at the extremes in order to bring the finite boundaries of the observable counts into coincidence with the infinite extremes of the implied measures.

The monotonicity between count and measure hold only when data are complete with no adaptive tailoring of item difficulties to person abilities. Only after interval linear measures have been successfully constructed does it become reasonable to proceed with statistical analysis in order to determine the predictive validity of measures or to compare measures.

Counts and Measures

Counts are made of concrete, tangible objects or visible events. The objects counted must be interpreted to be exchangeable, as though they were identical, for the purpose of counting. However, in their concreteness they are manifestly not identical. Upon sufficient examination, objects counted are always uniquely different.

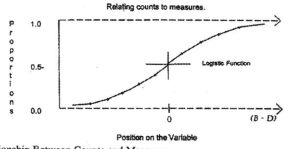

Figure 3. The Relationship Between Counts and Measures

Thus counting is a daring but necessary fiction. The fiction necessary for counting different events as "the same thing again" prompts us to treat each object as "the same thing" again, even though we know that no two objects encountered can be exactly the same. Without the fiction of "same again," counting cannot proceed. Since counts are built on fiction, the invention and governance of what to count requires an observation model. (See Vaihinger ([1925] 1935) for the history and application of "fictions" to the arts and sciences.) Once we acknowledge that concrete objects are fictitious in our counting, we reconsider their actual concrete differences as clumsy and think it better to "count" perfect ideas rather than clumsy objects. Our goal is "objects" that are ideally equal, exchangeable units. The observation model is the means by which we define what to count. The measurement model is the method by which we construct estimates of ideal magnitudes that are the inventions of measurement from crude concrete counting.

Reliability

A reliability coefficient is commonly used to decide how well a variable correlates with itself over independent applications. It is composed of pointbiserial correlations between item responses and total scores. The traditional reliability coefficient is a mixture of item fit and measure precision, person fit, and spread. Rasch measurement separates these parts into the "sample" standard deviation and the "test" root mean square error. Measurement precision depends on the number of items administered, the extent to which the items are on target, and the degree to which the respondent uses the items coherently. Person spread is a sample characteristic.

Main considerations in any measurement application are the standard errors of measurement. Rasch measurement produces a measure of each person's ability on a linear scale with an accompanying standard error. This individualization of measurement error actually obviates any need for considering a global estimate of sample specific reliability for person measure precision. The first question is, how many items are needed in a given region of the variable to obtain a measure precise enough to meet the current need for precision in that region?

A separation index is also available. Separation is the ratio of an unbiased, error-corrected, estimate of the sample standard deviation to the root mean square measurement error of the sample. It quantifies "reliability" in a simple and direct way with a clear interpretation.

Person separation "reliability" is comparable to the KR20 measure of internal consistency. Corrected for degrees of freedom [L / (L -1)], the Rasch measurement Person Separation Reliability (PSR) yields a result algebraically similar index to KR20:

$$KR20= (L - L - 1) (1 - \sum pq /\sigma_y^2$$

and

$$Rasch\ PSR= (L - L - 1) (1 - MSE_p/SD_p^2$$

In these formulae:

pq is the error variance of an item for which the sample item p-values apply,

s_y^2 is the sample variance of the nonlinear person raw scores,
MSE_p is the sample average person measure error variance in logits,

SD_p^2 is the sample variance of the linear logit person measures,

and

PSR is the Rasch person separation ratio.

KR20 is commonly calculated for items but almost never for persons. Hoyt (1941) in his paper on test reliability by ANOVA recognized that "extended examination of the 'among' items variance would make it possible to decide on the heterogeneity of the respective difficulties of the items" (156). Good advice that has been rarely practiced.

The Rasch measurement, Item Separation Reliability (ISR), routinely calculated by WINSTEPS (Linacre 2002), indicates how well items are separated by persons taking the test.

That formula is

$$ISR = 1 - (MSE_I / SD_I^2)$$

MSE_I is the test average item calibration error variance in logits, and SD_I^2 is the test variance of the linear logit item calibrations.

Validity

Qualitative validity refers to the abstract idea of the variable. The conceptual plan of a variable is its qualitative validity. Content validity is demonstrated by the calibrations of items planned and written to bring the variable to life. Successful item writing begins with thinking of the variable as a line with direction (an arrow) and arranging items according to their intended difficulty along this line. This lays out a substantive definition of a single dimension and requires empirical confirmation. A dialogue results between the tasks of writing items according to their intended difficulties and gathering empirical evidence supporting these intentions. This is the process of building content validity.

When such a plan and its resultant items provide a useful representation of a variable, the items define the variable operationally, and it is manifest by responses to the items. If the structure of the variable is supported by the item calibrations, we have construct validity.

The quantitative aspects of validity are commonly called "criterion" and "predictive" validity according to the Standards for Educational and Psychological Tests (APA 1999). These forms of validity are usually expressed by correlation coefficients. Rasch measurement helps us to see that there are only two types of validity that can be evaluated from item response data: (1) the ordering of items and persons and (2) the fit of items and persons.

Order validity derives its meaning from the calibration order of items and embodies traditional content and construct validity. Utility validity derives from the measurement order of person characteristics and so embodies criterion validity.

Fit validity is determined in several ways. It is determined for every response by the discrepancy of the response from its expectation. This identifies "useful" items and measures. Item function validity is determined by an analysis of a sample of responses to an item in order to determine which items are not working the way they were intended. Person performance validity is determined by an analysis of the items responded to by a person. This identifies persons not responding in the way we expect.

Smith (2001) reviews the essential issues of Rasch reliability and validity. More details on the derivation of item and person fit statistics, power, Type I error rates and distribution properties are given in Smith (1991a, 1991b).

Sample Issues

The application of Rasch measurement to instrument construction has a direct connection with sampling. The resulting estimates of the item parameters are independent of the sample distribution of person measures. This is a specific consequence of the conjoint additivity of person ability B and item difficulty D. If we have two person abilities B_1 and B_2 for a single item difficulty D_5, and subtract them, that is, $(B_1 - D_5) - (B_2 - D_5)$, item difficulty D_5 cancels leaving only the ability difference $(B_1 - B_2)$ of the two persons as the remainder. See Rasch (1977, 63–66) and Rasch ([1960] 1980; 1994, 18–21) for more details.

The consequences of this simple example generalize. We can estimate differences between person measures regardless of which items are used and estimate differences between item calibrations regardless of which persons provide the data. If it were not so, we would not have measurement. When data do not serve this requirement, we do not have measurement.

The consequences are remarkable. When items and persons are reasonably targeted and we estimate an item difficulty D, the actual distribution of the ability B's does not matter. And likewise, when we estimate person ability B, the distribution of the D's does not matter. We can estimate a statistically equivalent B from any subset of targeted item difficulty D's. Likewise, we can estimate a statistically equivalent D from any subset of targeted person B's.

This estimation process can be demonstrated and verified by dividing the sample into any two groups, in particular, high scorers and low scorers. We estimate D twice, once from each sample. The estimates of D will be statistically equivalent given the error of measurement. This is the essential property of the conjoint additive Rasch model.

The application of this estimation process, however, requires that any sample of persons chosen be on target for the instrument. If all persons answer all items correct, or answer no items correct, we cannot make estimates of person measures or item difficulties. The utility of conjoint additivity requires that the instrument match the persons for whom it is intended. When this requirement is met, the particular local distribution of person abilities among the sample does not affect estimates of item difficulties.

If, however, we administer an instrument, spelling, for example, to two different samples such as those who learned English as their first language and those who have learned English as a second language, we may get different sets of item difficulty estimates. In such instances the differences are the consequence of an interaction between the substantive differences of the person samples and item text. The resulting item difficulties are not free from the differing language properties of the person samples.

Persons who take a test have to be exchangeable with respect to the measure. If this did not occur, then every value would be unique and no generalizations could occur. What this means is that the test has to approximate the same meaning for all persons. When this is not the case, then we can expect to encounter unpredictable responses. Native English spellers may differ from nonnative English spellers. If more than one dimension exists, exchangeability cannot be assured. If there is a second dimension differentially observed among the persons and items, then exchangeability does not exist. Consider measuring height. If "height" were uniquely defined for each person, then we could not have the abstract idea of measuring height. To measure height means that we consider this single dimension to be free of such characteristics as name, country of origin, and weight. Height is a single dimension that demonstrates an exchangeable property of people.

Consider currency. You have ten one-dollar bills in your wallet. Any one of them is exchangeable for another in making a purchase. Any three will do, if the price is three dollars. This is because your dollars are exchangeable. Some dollars may differ by other characteristics such as wrinkles versus crisp, clean versus dirty. But this constitutes a different attribute. What make commerce work is that a valid dollar is exchangeable with any other valid dollar. The abstract concept of "dollar" can be substantiated by any valid, exchangeable instance of the same. This holds true to any concept operationally defined by a single dimension of interest.

Summary

Variable construction requires careful attention to substantive issues, and the theory that should guide it's development. The resulting observations require monitoring before measurement can be assumed. Rasch measurement practitioners should give careful attention to these matters because subsequent data analysis can only be successful when proper consideration has been paid to these prerequisite issues.

Box 22: Combined Gas Law and a Rasch Reading Law

Many physical laws are expressed as universal conditionals among variable triplets. Newton's second law, for example, formalizes the relationship between mass and acceleration when holding force constant (i.e., conditioning on) as $F = MA$. Similarly, the combined gas law specifies the relationship between volume and temperature conditioning on pressure. After transformation (e.g., $\log_e$ pressure + $\log_e$ volume - $\log_e$ temperature = constant, given a frame of reference specified by the number of molecules), each of these laws can be abstracted to a common form (a + b - c = constant). Note that these laws permit causal claims expressible as counterfactual conditionals. If we have twenty liters of a gas at 2000°K under twenty atmospheres of pressure and we cool the gas to 1000°K, we will observe a decrease in the pressure to ten atmospheres.

The value of such laws may lie more in the explicit causal organization of key constructs than in accuracy of prediction in the real world.

212

Cartwright (1983) made a useful distinction between "the tidy and simple mathematical equations of abstract theory, and the intricate and messy descriptions, in either words or formulae, which express our knowledge of what happens in real systems made of real materials" (128). This distinction led Cartwright to the view that "fundamental equations do not govern objects in reality; they govern only objects in models [i.e., idealizations]" (129).

The human sciences, for the most part, lack laws such as those stated above and consequently lack causal stories that are universal in application: "Lacking a 'complete (causal) theory' of what influences what, and how much, we simply cannot compute expected numerical changes in stochastic dependencies when moving from one population or setting to another" (Meehl 1978, 814, emphasis in original). In this note, we build on the abstracted formalism derived above and imagine the form of a Rasch Reading Law.

Table 1

Comprehension Rates for Readers of Different Ability with Texts of the Same Text Complexity or How Reader Ability and Comprehension Rate Relate Under Constant Text Complexity

ReaderAbility	Sports Illustrated Complexity	Comprehension Rate
500L	1000L	25%
750L	1000L	50%
1000L	1000L	75%
1250L	1000L	90%
1500L	1000L	96%

Contemporary reading theory recognizes three related constructs: reader ability (a stable attribute of persons), text complexity (a stable attribute of text), and comprehension (the rate at which a particular reader makes meaning from a particular text). As a result of 25 years of ongoing research, we know that comprehension is a function of the difference between reader ability and text complexity (Stenner & Burdick, 1997). Table 1 illustrates the relationship between reader ability and comprehension rate with text complexity held constant. With increasing reader ability, the model forecasts increasing comprehension rate conditioning on text complexity. This description of the relationship between reader, text, and comprehension echoes the description of the combined gas law (Table 2).

Table 2 How temperature and pressure relate under constant volume		
Temperature Volume Pressure		
2000°K	20 Liters	20.0 atm
1000°K	20 Liters	10.0 atm
500°K	20 Liters	5.0 atm
250°K	20 Liters	2.5 atm
125°K	20 Liters	1.25 atm

In fact, logit transformed comprehension rate + text measure - reader measure = the constant 1.1 (given a frame of reference that specifies 75 percent comprehension whenever text measure = reader measure). Therefore, a + b - c = constant holds as the common abstracted form of both the combined gas law and the Rasch Reading Law as well as many other physical laws. Below are several causal corollaries of the Rasch Reading Law.

(1) For any reader (and thus for all readers), an increase in text measure causes a decrease in comprehension.
(2) For any reader (and thus for all readers), a decrease in text measure causes an increase in comprehension.
(3) For any text (and thus for all texts), an increase in reader ability causes an increase in comprehension.
(4) For any text (and thus for all texts), a decrease in reader ability causes a decrease in comprehension.

Corollaries such as those above are consequences of the highly abstracted a + b - c = constant, holding in a domain of enquiry. Tables 1 and 2 concretize this abstraction for the gas law and reading law. The Rasch model, in concert with a substantive theory, is a powerful tool for discovering and testing the adequacy of such formulations. Note, however, that the fact that data fit a Rasch model says nothing about causality. Rasch models are associational rather than causal. Substantive theory provides the causal story for the variation detected by a measurement procedure. Specification equations formalize these causal stories and allow precise predictions. "These causal explanations have truth built into them. When I infer from an effect to a cause, I am asking what made the effect occur, what brought it about. No explanation of that sort explains at all unless it does present a cause, and in accepting such an explanation, I am accepting not only that it explains in the sense of organizing and making plain, but also that it presents me with a cause" (Cartwright 1983, 91).

If one of our children cannot summarize well what he just read in his fifth grade science text, we explain this by pointing out that he is a 580L reader and the textbook is at 830L. The equation that models comprehension rate as a function of the difference between reader measure and text measure produces an expected comprehension rate below 50 percent. We hypothesize that the child's failure to produce a good summary has a cause: low comprehension. Suppose that we go to the Web and find a 600L article on the same science topic, and the child reads the article and produces a coherent summary of the text. We conclude that, indeed, low comprehension was the cause of poor summarization. Manipulating the reader-text match caused an increase in comprehension, which in turn caused a change in summary performance. Clearly I am inferring from effect to probable cause. Note that this explanation is unintelligible "without the direct implication that there are [readers, texts and comprehension rates]" (Cartwright 1983, 92).

We wonder how many other variable triplets in the human sciences can be abstracted to the form a + b - c = constant. The implications of this kind of lawmaking for construct validity should be evident (see Borsboom 2005).

Box 23: Formative and Reflective Models

Structural equation modeling (SEM) distinguishes two measurement models: reflective and formative (Edwards and Bagozzi 2000). Figure 1 contrasts the very different causal structures hypothesized in the two models. In a reflective model (left panel), a latent variable (e.g., temperature, reading ability, or extraversion) is posited as the common cause of item or indicator behavior. The causal action flows from the latent variable to the indicators. Manipulation of the latent variable (via changing pressure, instruction, or therapy, for example) causes a change in indicator behavior. Contrariwise, direct manipulation of a particular indicator is not expected to have a causal effect on the latent variable.

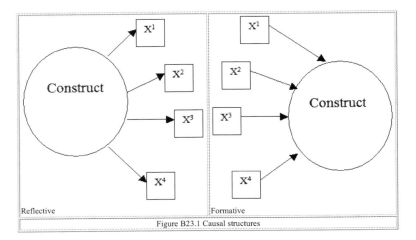

Figure B23.1 Causal structures

A formative model, illustrated on the right-hand side of figure B23.1, posits a composite variable that summarizes the common variation in a collection of indicators. A composite variable is considered to be composed of independent, albeit correlated, variables. The causal action flows from the independent variables (indicators) to the composite variable. As noted by Bollen and Lennox (1991), these two models are conceptually, substantively, and psychometrically different. We suggest that the distinction between these models requires a careful consideration of the basis for inferring the direction of causal flow between the construct and its indicators.

Given the primacy of the causal story we tell about indicators and constructs, what kind of experiment, data, or analysis could differentiate between a latent variable story and a composite variable story? For example, does a Rasch analysis or a variable map or a set of fit statistics distinguish between these two different kinds of constructs? We think not! A Rasch model is an associational (think: correlational) model and as such is incapable of distinguishing between the latent-variable-causes-indicators story and the indicators-cause-composite-variable story.

Some examples from without and within the Rasch literature should help illustrate the distinction between formative and reflective models. The paradigmatic example of a formative or composite variable is socioeconomic status (SES). Suppose the four indicators are education, occupational prestige, income, and neighborhood. Clearly, these indicators are the causes of SES rather than the reverse. If a person finishes four years of college, SES increases even if where the person lives, how much they earn, and their occupation stay the same. The causal flow is from indicators to construct because an increase in SES (job promotion) does not imply a simultaneous change in the other indicators. Bollen and Lennox

(1991) gave another example: life stress. The four indicators are job loss, divorce, recent bodily injury, and death in the family. These indicators cause life stress. Change in life stress does not imply a uniform change in probabilities across the indicators. Lastly, the construct could be accuracy of eyewitness identification, and its indicators could be recall of specific characteristics of the person of interest. These characteristics might include weight, hairstyle, eye color, clothing, facial hair, voice timber, and so on. Again, these indicators cause accuracy; they are not caused by changes in the probability of correct identification.

The examples of formative models presented above are drawn from the traditional classical test theory (CTT), factor analysis, and SEM literatures. Are Rasch analyses immune to confusion of formative and reflective models?

Imagine constructing a reading rating scale. A teacher might complete the rating scale at the beginning of the school year for each student in the class. Example items (rating structure) might include: (1) free or reduced price lunch (1,0), (2) periodicals in the home (0,1,2,3), (3) daily newspaper delivered at home, (4) student read a book for fun during the previous summer (1,0), (5) student placement in reading group (0,1,2,3), (6) student repeated a grade (1,0), (7) students current grade (1,2,3, …), (8) English is student's first language (1,0), and so on. Now, suppose that each student, in addition to being rated by the teacher, took a Lexile-calibrated reading test. The rating scale items and reading test items could be jointly analyzed using WINSTEPS or RUMM2020. The analysis could be anchored so that all item calibrations for the reading rating items would be denominated in Lexiles. After calibration, the best-fitting rating scale items might be organized into a final scale and accompanied by a scoring guide that converts raw counts on the rating scale into Lexile reader measures. The reading scale is conceptually a composite formative model. The causal action flows from the indicators to the construct. Arbitrary removal of two or three of the rating items could have a disastrous effect on the predictive power of the set and, thus, on the very definition of the construct, whereas, removal of two or three reading items from a reading test will not alter the construct's definition. Indicators (e.g., items) are exchangeable in the reflective case and definitional in the formative case.

Perline, Wainer, and Wright (1979), in a classic paper, used parole data to measure a latent trait which might be labeled "the ability to successfully complete parole without any violations" (235). Nine dichotomously scored items rated for each of 490 participants were submitted to a BICAL analysis. The items were rated for presence or absence of: high school diploma or GED, eighteen years or older at first incarceration, two or fewer prior convictions, no history of opiate or barbiturate use, release plan to live with spouse or children, and so on. The authors concluded, "In summary, the parole data appeared to fit [the Rasch Model] overall ... However, when the specific test for item stability over score groups was performed ... there were serious signs of item instability" (249). For our purposes, we simply note that the Rasch analysis was interpreted as indicating a latent variable when it seems clear that it is likely a composite or formative construct.

A typical Rasch analysis carries no implication of manipulation and thus can make no claim about causal action. This means that there may be little information in a traditional Rasch analysis that speaks to whether the discovered regularity in the data is best characterized as reflective (latent variable) or formative (composite variable).

Rasch models are associational (i.e., correlational) models, and because correlation is necessary but not sufficient for causation, a Rasch analysis cannot distinguish between composite and latent variable models. As the Rubin-Holland framework for causal inference specifies: no causation without manipulation. It seems that many Rasch calibration efforts omit the crucial last step in a latent variable argument, that is, answering the question "What causes the variation that the measurement instrument detects?" (Borsboom 2005). We suggest that there is no single piece of evidence more important to a construct's definition than the causal relationship between the construct and its indicators.

Box 24: Indexing versus Measuring

In RMT 22:1, Stenner, Stone, and Burdick (2008) distinguished between two different measurement models: reflective or latent variable models, and formative or composite variable models (Edwards and Bagozzi 2000). In the former, the causal action flows from the latent variable to the indicators (e.g., temperature), whereas in the latter, the causal action flows from indicators to the composite variable (e.g., socioeconomic status). We believe that the language we use should accentuate these differences, and as such, we propose to call reflective models "measurement models"; what these models produce we will call measures, and the process of producing these measures will be called measuring. In parallel fashion, formative models will be called index models, what they produce we will call indices, and the process of producing indices will be called indexing. The notion of an index is well developed in economics and sociology and carries the connotations we desire. What follows is a discussion of how indexing and measuring differ and why it is important to make this distinction in the human sciences.

Definitions

Indices are the effects of their indicators, whereas measures (of latent variables) are the causes of their indicators. So, differences in stature or consumer price behavior are caused by changes in height (or weight) and price changes for commodities (computers, milk, gasoline), respectively. Changes in latent variable measures, in contrast, cause a homogeneous (often nonlinear) change in indicator behavior, as when temperature change causes thermometric fluid to expand in the thermometer or a change in reader ability causes a change in count correct on a reading test.

Altering the indicators of an index changes the definition of the variable being indexed, whereas changing the indicators for a measure will not alter the latent variable (although precision of measurement and (or) unit size may be affected). So, if midline girth is added to height and weight as indicators of stature, or all electronic commodities are eliminated from the Consumer-Product-Index (CPI) market basket, the definition of what is being indexed changes.

In contrast, knowledge of expansion coefficients and viscosity differences allows us to swap new thermometric fluids for mercury without changing the construct being measured. Similarly, new reading items with different text and item types can be swapped for previous items without changing the construct being measured.

Another way to express this point is that the indicators for an index are constitutive of that index, whereas indicators for a latent variable are incidental to the construct's definition.

Specification Equations

In a generally objective measurement framework (e.g., Lexiles), what is crucial in the definition of the construct is the specification equation that specifies the cause of the variation detected by the instrument. Because the indicators of an index by design track different kinds of variation (height, weight, midline girth), it is difficult to imagine a specification equation that could, somehow, capture what these indicators share independent of the linear (or otherwise) combination that constitutes the index. What, for example, would a parallel form of Sheldon's somatotype rating scale look like? Difficulty in imagining what new indicators would constitute a parallel form is strongly suggestive of the need for an index rather than a measurement model.

Indices Misinterpreted as Latent Variables

Because both index and measurement models are fundamentally associational (i.e., based on correlations among indicators), traditional applications of Rasch model software often cannot distinguish between an index and a latent variable (Stenner, Burdick, and Stone 2008). Examples of resulting confusion take predominantly one particular form—index variables are interpreted as if they are latent variables. Here is an example typical of the Rasch literature [and RMT, Ed.]:

> The Rasch model has been shown to fit FIM data reasonably well, which indicates that the scale locations describe adequately the relative order in which these functions are lost in the aging population. The items on the top describe difficult activities, such as climbing stairs, whereas items on the bottom describe easier activities that are maintained relatively well. (Embretson 2006, 52)

Contrary to a latent variable interpretation, the FIM (Functional Independence Measure) appears to be an index of motor functioning with the causal action moving from indicators to index. If the desired medical outcome is "more functional independence," then rehabilitating bladder control, walking, bathing, and so on should promote the intended outcome rather than the other way around. Alternatively, we could teach the patient to drive a motorized wheelchair, but to include this ability as an indicator would alter the definition of functional independence.

Global fit of data to a Rasch model will not resolve the direction of causal flow and thus will not provide unambiguous evidence for a latent variable interpretation of the construct. A substantive theory and associated specification equation capable of explaining variation in indicator difficulties is a big step in support of a latent variable interpretation. The coup de grace is a demonstration of the specification equation's causal status using experimental manipulation of instrument characteristics (radicals) and subsequent observation of the theoretically predicted change in the measurement outcome.

Correlation Is Not Causation

It is a property of indices (economical, sociological, or psychological) that the indicator composite may be found to correlate with an unintended criterion more strongly than the intended one. Such a discomforting outcome is yet another reason that a correlational (as opposed to a causal) view of validity is not sustainable.

Latent variable interpretations are most defensible when global fit of data to a Rasch model is accompanied by invariance of the indicator structure throughout the range of the construct. In the language of additive conjoint measurement (Luce and Tukey 1964) and as realized in the Lexile Framework for Reading (Kyngdon 2008b), it should be possible to trade off a difference between reader abilities of 200L for a difference in text complexity of 200L to hold comprehension rate (count correct / total items) constant (Burdick, Stone, and Stenner 2006). This trade-off property has been shown to operate throughout the grade range from kindergarten to advanced adult reading (e.g., Supreme Court decisions) and would not be expected to hold for a reading index variable composed of items such as (1) number of books in the home; (2) daily newspaper subscription; (3) English as a first language; and so on.

It may be true that "where there is correlational smoke there is likely to be causational fire" (Holland 1986, 951). Good fit with a Rasch model is correlational smoke, but as we have just seen, it takes an experimental test of a substantive theory to unambiguously distinguish between a latent variable and an index.

THE CUBIT: A HISTORY AND MEASUREMENT COMMENTARY

8

Introduction

Much of what we know today about the cubit comes from Hebrew scripture and the Old and New Testaments. Many people have heard or read about the dimensions of Noah's ark or Solomon's Temple. Egyptian history is another common site to encounter the "cubit," from the dimensions given for pyramids and temples. The cubit was a common unit in the early East. It continues today in some locations but with less prominence, having been replaced by modern-day units. Early employment of the cubit throughout the Near East showed varied dimensions for this unit. Some variants can be examined more easily with reference to biblical passages. Additional variants can also be found in secular documents, but these are less known and less accessible than scripture.

The word cubit ('kyü-bət) in English appears to be derived from the Latin *cubitum*, meaning "elbow" (πῆχυς (pay'-kus) in Greek). The cubit is based upon a human characteristic—the length of the forearm from the tip of the middle finger to end of the elbow. Many definitions seem to agree on this aspect of the unit, yet it does not produce a universal standard given that there are many ways to determine a cubit. It can be measured from the elbow to the base of the hand, from the elbow to a distance located between the outstretched thumb and little finger, or from the elbow to the tip of the middle finger. These alternate descriptions further complicate the matter of determining a specific unit measure of the cubit. Hereafter, the latter description, elbow to the tip of the middle finger, will signify the common unit.

The human figure (typically male) has been the basis for many dimensions. The foot is immediately recognized as an example. Less commonly heard is onyx (nail), which remains a medical term. The Old English *ynche*, *ynch*, *unce*, or *inch* was a thumb-joint breadth. The anthropomorphic basis for many standards supports the statement "man is the measure of all things" attributed to Protagoras according to Plato in the *Theaetetus*. Small wonder the cubit was initially employed for measurement given its omnipresent availability for use. Human figure units, which are arbitrary but universal, are especially effective as their bodily reference produces a crude standard that is immediately accessible.

The cubit provides a convenient middle unit between the foot and the yard. The English yard, which could be considered a double cubit, is said to measure twelve palms (about 90 cm, or thirty-six inches) and is measured from the center of a man's body to the tip of the fingers of an outstretched arm. This is a useful way of measuring cloth held center body to an outstretched hand (two cubits), or across the body to both outstretched hands (four cubits as specified in Exodus 26:1–2, 7–8). The English ell is a larger variant of the cubit consisting of fifteen palms (114 cm, or forty-five inches). It is about equal to the cloth measure *ell* of early Scotland. A man's stride, defined as stepping left-right, produces a double cubit, or approximately a yard.

The dimensions in table 1 give the approximate relative lengths for meter, yard, cubit, and foot.

TABLE 1: The relative lengths of four common dimensions.

Meter	
Yard	
Cubit	
Foot	

Table 8.1 The relative lengths of four common dimensions

The cubit was a basic unit in early Israel and the surrounding Near East countries. It is אמה in Hebrew (pronounced am-mah'), which can be interpreted as "the mother of the arm" or the origin—that is, the forearm/cubit. Selected biblical references for the cubit include these five rather well-known selections:

> (1) And God said to Noah, I have determined to make an end of all flesh; for the earth is filled with violence through them; behold, I will destroy them with the earth. Make yourself an ark of gopher wood; make rooms in the ark, and cover it inside and out with pitch. This is how you are to make it: the length of the ark three hundred cubits, its breadth fifty cubits, and its height thirty cubits. (Genesis 6:13–15 RSV)

(2) They shall make an ark of acacia wood; two cubits and a half shall be its length, a cubit and a half its breadth, and a cubit and a half its height. And you shall overlay it with pure gold, within and without shall you overlay it, and you shall make upon it a molding of gold round about. (Exodus 25:10–11 RSV)

(3) And he made the court; for the south side the hangings of the court were of fine twined linen, a hundred cubits; their pillars were twenty and their bases twenty, of bronze, but the hooks of the pillars and their fillets were of silver. And for the north side a hundred cubits, their pillars twenty, their bases twenty, of bronze, but the hooks of the pillars and their fillets were of silver. And for the west side were hangings of fifty cubits, their pillars ten, and their sockets ten; the hooks of the pillars and their fillets were of silver. And for the front to the east, fifty cubits. (Exodus 38:9–13 RSV)

(4) And Saul and the men of Israel were gathered, and encamped in the valley of Elah, and drew up in line of battle against the Philistines. And the Philistines stood on the mountain on the one side, and Israel stood on the mountain on the other side, with a valley between them. And there came out from the camp of the Philistines a champion named Goliath, of Gath, whose height was six cubits and a span. (1 Samuel 17:2–4 RSV)

(5) In the four hundred and eightieth year after the people of Israel came out of the land of Egypt, in the fourth year of Solomon's reign over Israel, in the month of Ziv, which is the second month, he began to build the house of The Lord. The house which King Solomon built for The Lord was sixty cubits long, twenty cubits wide, and thirty cubits high. (1 Kings 6:1–2 RSV)

The cubit determined a measure for many aspects of life in biblical history. A Sabbath day's journey measured two thousand cubits (Exodus 16:29). This statute proscribed a limit to travel on the Sabbath. The distance between the Ark of the Covenant and the camp of the Israelites during the Exodus is estimated at about 914 meters, a thousand yards, or two thousand cubits.

Biblical citations and historical archeology suggest more than one standard length for the cubit existed in Israel. A citation in 2 Chronicles 3:3 may imply cubits of the old standard, while Ezekiel 40:5 and 43:13 may be indicating the cubit plus a hand. Archeological evidence from Israel suggests that 52.5 cm (20.67 inches) and 45 cm (17.71 inches) constitute the long and short cubits of this time and location. To some scholars, the Egyptian cubit was the standard measure of length in the biblical period. The biblical sojourn/exodus, war, and trade are probable reasons for this length to have been employed elsewhere.

The Tabernacle, the Temple of Solomon, and many other structures are described in the Bible by cubit measures. These also occur with two different cubit dimensions, the long or royal (architectural) cubit and the short (anthropological) cubit. Scholars have used various means to determine the length of these cubits with some success. The long cubit is given as approximately 52.5 centimeters, and the short cubit as about forty-five centimeters.

The Israelite long cubit corresponds to the Egyptian cubit of seven hands, with six hands for the shorter one. *Eerdman's Dictionary of the Bible* states "... archeology and literature suggests an average length for the common cubit of 44.5 cm (17.5 in.)." This citation also gives a range of 42–48 cm (17–19 in) for the cubit. Range is an important parameter because it indicates the variation operating on this measure, which in this case indicates multiple influences.

The English use of *cubit* is difficult to determine. The exact length of this measure varies, depending upon whether it includes the entire length from the elbow to the tip of the longest finger or one of the alternate reference points described earlier. Some scholars suggest that the longer dimension was the original cubit and conclude that the correct measurements are 20.24 inches for the ordinary cubit and 21.88 inches for the sacred one, or a standard cubit from the elbow to end of middle finger (20") and a lower forearm cubit from the elbow to the base of the hand (12"). These dimensions correspond to the Egyptian measurements in Easton's *Illustrated Bible Dictionary. The Interpreter's Bible* gives the Common Scale length as 444.25 mm or 17.49 inches and Ezekiel's Scale as 518.29 mm or 20.405 inches for the two cubit lengths. Inasmuch as the Romans colonized England, the shorter cubit previously mentioned may have been the standard.

A rod or staff is called דמג (*gomedh*) in Judges 3:16, which means "a cut," or "something cut off." The LXX (Septuagint) and Vulgate render it "span," which in Hebrew scripture or the Old Testament is defined as roughly eighteen inches (the forearm cubit). Among the several cubits mentioned is the cubit of a man or common cubit in Deuteronomy 3:11 and the legal cubit or cubit of the sanctuary described in Ezra 40.5.

Barrios gives a summary of linear Hebrew measures (see table 2).

TABLE 2: Hebrew linear measures.

Measure	Common scale		Ezekiel's scale	
	Millimeters	Inches	Millimeters	Inches
Cubit	444.25	17.49	518.29	20.405
Span	222.12	8.745	259.14	10.202
Handbreadth	74.04	2.91	74.04	2.91
Finger	18.51	0.72	18.51	0.72

Table 8.2 Hebrew linear measures

Barrois indicates the dimension of the cubit can only be determined by deduction and not directly because of conflicting information. He reports the aqueduct of Hezekiah was 1,200 cubits according to the inscription of Siloam. Its length is given as 5333.1 meters or 1,749 feet. Absolute certainty for the length of a cubit cannot be determined, and there are great differences of opinion about this length fostering strong objections and debates. Some writers make the cubit eighteen inches, and others twenty, twenty-one inches, or greater. This appears critically important for those seeking to determine the exact modern equivalent of dimensions taken from scripture. Taking twenty-one inches for the cubit, the ark Noah built would be 525 feet in length, 87 feet 6 inches in breadth, and 52 feet 6 inches in height. Using the standard 20" cubit and 9" span, Goliath's height of six cubits plus a span would be 10 feet and 9 inches. With a cubit of 18" his height is 9 feet 9 inches. The Septuagint, LXX, suggests four cubits plus a span, or a more modest 6 feet 9 inches. There are many implications depending upon which dimension is selected. The intended effect of the story depends on young David slaying a giant, and not simply an above-average sized man! Likewise, many other dimensions and descriptions in early writings use the larger dimensions for a better story. Sacred dimensions require solemn, awe-inspiring descriptions, but this frustrates an exact determination.

Rabbi David ben Zimra (1461–1571) claimed the *Foundation Stone* and *Holy of Holies* were located within the Dome of the Rock on the Temple Mount. This view is widely accepted but with differences of opinion over the exact location known as the "central location theory," some of these differences result from strong disagreement over the dimension of the cubit. Kaufman argues against the "central location theory" defending a cubit measuring 0.437 meters (1.43 feet). David argues for a Temple cubit of 0.56 meters (1.84 feet).

Differences in the length of the cubit arise from various historical times and geographical locations in the biblical period. These very long time periods and varied geographical locations frustrate assigning a more exact measurement to the cubit. Israel's location between Egypt and Mesopotamia suggests that many influences came into play over the space of hundreds of years in this well-traveled area. These influences likely contributed to the wide variety of dimensions we encounter across this long span of time.

The earliest written record of the cubit occurs in the Epic of Gilgamesh. This incomplete text is extant in twelve tablets written in Akkadian, and found at Nineveh in the library of Ashurbanipal, king of Assyria (669–630? BCE). Other fragments that date to 1800 BCE contain parts of the text, and still more fragments mentioning this epic have been found dating from the second millennium BCE. The cubit is specifically mentioned in the text when describing a flood as remarkably similar to, while predating, the biblical flood in Genesis. Obviously, the cubit was an early and important unit of the Middle East, fundamental to conveying linear measures as shown in tables 2, 3, and 4.

TABLE 3

Great Pyramid at Gizeh, Khufu	20.620 ± -.005
Second Khafra	20.64 ± -.03
Granite temple	20.68 ± -.02
Third Pyramid Menkaura	20.71 ± .02
Peribolus walls	20.69 ± -.02
Great Pyramid of Dahshur (?)	20.58 ± -.02
Pyramid at Sakkara Pepi	20.51 ± -.02
Fourth to sixth dynasty, mean of all	20.63 ± -.02

Table 8.3

TABLE 4

Egyptian common cubit	18.24 inches
Egyptian royal cubit	20.64 inches
Great Assyrian cubit	25.26 inches
Beládi cubit	21.88 inches
Black cubit	20.28 inches

Table 8.4

Egypt

The Egyptian hieroglyph for the cubit shows the symbol of a forearm. However, the Egyptian cubit was longer than a typical forearm. It seems to have been composed of seven palms of four digits each, totaling twenty-eight parts, and was about 52.3–52.4 cm in length according to Arnold.

The earliest accepted standard measure is from the Old Kingdom pyramids of Egypt. It was the royal cubit (*mahe*). The royal cubit was 523 to 525 mm (20.6 to 20.64 inches) in length and was subdivided into seven palms of four digits each, for a twenty-eight-part measure in total. The royal cubit is known from Old Kingdom architecture dating from at least as early as the construction of the Step Pyramid of Djoser around 2700 BCE.

Petrie begins chapter XX of *Values of the Cubit and Digit* writing,

> *The measurements which have been detailed in the foregoing pages supply materials for an accurate determination of the Egyptian cubit. From such a mass of exact measures, not only may the earliest value of the cubit be ascertained, but also the extent of its variations as employed by different architects.*

Petrie's methods and findings are so clearly and precisely described they can best be quoted as follows:

> *For the value of the usual cubit, undoubtedly the most important source is the Kings Chamber in the Great Pyramid; that is the most accurately wrought, the best preserved, and the most exactly measured, of all the data that are known.*

Arranging the examples chronologically, the cubit used was as shown in table 8.3.

Petrie writes the following.

> *For the cubit I had deduced from a quantity of material, good, bad, and indifferent, 20–64 .02 as the best result that I could get; about a dozen of the actual cubit rods that are known yield 20–65 .01; and now from the earliest monuments we find that the cubit first used is 20–62, and the mean value from the seven buildings named is 20–63 b .02-. ... On the whole we may take 20– as the original value and reckon that it slightly increased on an average by repeated copyings in course of time. (178–179)*

Greek and Roman Comparisons

In the writings of Eratosthenes, the Greek σχο νος (schoe'nus) was 12,000 royal cubits assuming a 0.525 meter. The stade was three hundred royal cubits (157.5 meters or 516.73 feet). Eratosthenes gave 250,000 stadia for circumference of the earth. Strabo and Pliny indicated 252,000 stadia for the circumference and seven hundred stadia for a degree. Reports of Egyptian construction indicate only a 0.04-inch difference between cubit of Snefru and Khufu pyramids according to Arnold and Gillings.

Lelgemann reported the investigation of nearly 870 metrological yardsticks whose lengths represent thirty different units. He argues for the earliest unit, the Nippur cubit, to be 518.5 mm. Lelgemann gives the ancient *stadion* = six hundred feet and reports the *stadion* at Olympia at 192.27 meters, which he believes is based on the Remen or old Egyptian trade cubit derived from the Egyptian royal cubit (523.75 mm) and old trade cubit = 448.9 mm.

Nichholson (1912), in *Men and Measures,* devoted a chapter to *The story of the cubit.* His summary (page 30) provided comparative lengths to five cubits as shown in table 4.

Nichholson proposes a long history of the cubit beginning before the time of the Great Pyramid of Kufu c. 2600 BCE. He claims a measure of five hundred common cubits for the base side indicating only a six-inch difference from the base measure made by Flinders Petrie. He fixes the date of the royal cubit at about 4000 BCE. The great Assyrian cubit is dated c. 700 BCE. The Beládic cubit is dated c. 300 BCE. Nichholson fixes the Black cubit as fully realized at around the ninth century of this era, which suggests a parallel to the growth and spread of Islam. While his measures for these variants of the cubit appear to dovetail with some of the other estimates given in this paper, there are serious questions about the chronological sequence associated with these variants. Nichholson offers no evidence or support for this sequence. His estimates of the common and royal cubits conform to other estimates, but the other values are less conforming.

Greek/Roman Periods

The Greek πηχυζ (pay'-kus) was a twenty-four-digit cubit. The Cyrenaica cubit measured about 463.1 mm with the middle cubit about 474.2 mm, making them roughly 25/24 and 16/15 Roman cubits. Other Greek cubits based on different digit measures from other Greek city-states were also used. The Greek forty-digit-measure appears to correspond to the Latin gradus, the step, or half a pace.

This shows that the Greeks and Romans inherited the foot from the Egyptians. The Roman foot was divided into both twelve unciae (inches) and sixteen digits. The uncia was a twelfth part of the Roman foot or pes of 11.6 inches. One uncia was 2.46 cm or 0.97 of our inch. The cubitas was equal to 24 digiti or 17.4 inches. The Romans also introduced their mile of one thousand paces or double steps, with the pace being equal to five Roman feet. The Roman mile of five thousand feet was introduced into England during the occupation. Queen Elizabeth, who reigned from 1558 to 1603, changed the statute mile to 5280 feet or eight furlongs, with a furlong being forty rods of 5.5 yards each. The furlong continues today as a unit common in horse racing.

The introduction of the yard as a unit of length came later, but its origin is not definitely known. Some believe the origin was the double cubit. Whatever its origin, the early yard was divided by the binary method into two, four, eight, and sixteen parts called the half yard, span, finger, and nail. The yard is sometimes associated with the "gird" or circumference of a person's waist, or with the distance from the tip of the nose to the end of the thumb on the body of Henry I. Units were frequently "standardized" by reference to a royal figure.

The distance between thumb and outstretched finger to the elbow is a cubit, sometimes referred to as a "natural cubit" of about 1.5 feet. This standard seems to have been used in the Roman system of measures as well as in different Greek systems. The Roman ulna, a four-foot cubit (about 120 cm), was common in the empire. This length is the measure from a man's hip to the fingers of the outstretched opposite arm. The Roman cubitus is a six-palm cubit of about 444.5 mm about 17.49 inches.

Other Near East Dimensions

Over time and the span of the geographic areas of the Middle East, various cubits and variations on the cubit have been recorded: 6 palms = 24 digits, that is, ~45.0 cm or 18 inches (1.50 ft.); 7 palms = 28 digits, that is, ~52.5 cm or 21 inches (1.75 ft.); 8 palms = 32 digits, that is, ~60.0 cm or 24 inches (2.00 ft.); and 9 palms = 36 digits, that is, ~67.5 cm or 27 inches (2.25 ft.). Oates [22, page 186] writing of Mesopotamian archeology states "measures of length were based on the cubit or "elbow" (very approximately 0.5 m)."

The Histories of Herodotus described the walls surrounding the city of Babylon as "fifty royal cubits wide and two hundred high (the royal cubit is three inches longer than the ordinary cubit)." An accompanying note to the text provides the information given in parentheses, and the endnote reports these values as "exceedingly high," raising questions about the height of these walls, which would be well over three hundred feet high if the royal cubit of twenty inches is implied, or one hundred meters if the royal cubit is 50 cm. For comparison, the great pyramid of Khufu is listed as originally 146.59 meters. The credibility of Herodotus has often been questioned, and these dimensions might be suspect also or subject to the same exaggerations found elsewhere in his reporting.

In 1916, during the last years of Ottoman Empire and during WWI, the German Assyriologist Eckhard Unger found a copper-alloy bar during excavation at Nippur from c. 2650 BCE. He claimed it to be a measurement standard. This bar, irregular in shape and irregularly marked, was claimed to be a Sumerian cubit of about 518.5 mm or 20.4 inches. A thirty-digit cubit has been identified from the second millennium BCE with a digit length of about 17.28 mm (slightly more than 0.68 inch). The Arabic Hashimi cubit of about 650.2 mm (25.6 inches) is considered to measure two French feet. Since the established ratio between the French and English foot is about sixteen to fifteen, it produces the following ratios: 5 Hashimi cubits ≈ 10 French feet ≈ 128 English inches. Also, the length of 256 Roman cubits and the length of 175 Hashimi cubits are nearly equivalent.

The guard cubit (Arabic) measured about 555.6 mm; 5/4 of the Roman cubit producing 96 guard cubits ≈ 120 Roman cubits ≈ 175 English feet. The Arabic nil cubit (or black cubit) measured about 540.2 mm. Therefore 28 Greek digits of the Cyrenaica cubit 25/24 of a Roman foot or 308.7 mm, and 175 Roman cubits or 144 black cubits. The Mesopotamian cubit measured about 533.4 mm, 6/5 Roman cubit making 20 Mesopotamian cubits ≈ 24 Roman cubits ≈ 35 English feet. The Babylonian cubit (or cubit of Lagash) measured about 496.1 mm. A Babylonian trade cubit existed that was nine-tenths of the normal cubit, that is, 446.5 mm. The Babylonian cubit is 15/16 of the royal cubit making 160 Babylonian trade cubits ≈ 144 Babylonian cubits ≈ 135 Egyptian royal cubits. The Pergamon cubit 520.9 mm was 75/64 of the Roman cubit. The Salamis cubit 484.0 mm was 98/90 of the Roman cubit. The Persia cubit of about 500.1 mm was 9/8 of the Roman cubit and 9/10 of the guard cubit. Extending the geographic area still further produces more names and values for the cubit.

From the Encyclopedia Britannica section on Weights and Measures given in volume 23, the unit specifications for the Middle East cubit are shown in table 8.5.

TABLE 5: Middle East names and dimensions for the cubit and related measures.

	Egypt		
Digit, zebo	1/28 royal cubit	0.737″	18.7 mm
Palm, shep	1/7	2.947″	75 mm
Royal foot	2/3	13.95″	254 mm
Royal cubit	unit	20.62	524
Ater, skhoine	12,000 royal cubits	3.9 miles	6.3 km
	Hebrew		
Finger, ezba	1/24 cubit	0.74″	19 mm
Palm, tefah	4 fingers, 1/6 cubit	2.9″	75 mm
Span, zeret	3 palms, 1/2 cubit	8.8″	225 mm
Royal cubit	7/6 standard cubit	20.7	525 mm
Pace	2 cubits	35.4″	900 mm
Stadion	360 cubits	528″	162 meters
	Greek		
Palm	4 fingers	3.0″	77 mm
Span	12 fingers	9.1″	231
Cubit	24 fingers	18.2″	463 mm
Stade		604 feet	185 meter

Table 8.5 Middle East names and dimensions
for the cubit and related measures

From a table in A. E. Berriman's *Historical Metrology*, we find his summary of cubit standards in table 6.

TABLE 6: Cubit dimensions from Berriman [8].

Cubit	Inches	Meter
Roman	17.48	0.444
Egyptian (short)	17.72	0.450
Greek	18.23	0.463
Assyrian	19.45	0.494
Sumerian	19.76	0.502
Egyptian (royal)	20.62	0.524
Talmudist	21.85	0.555
Palestinian	25.24	0.641

Table 8.6 Cubit dimensions from Berriman

If one assumes the values from Berriman's table to be reasonable estimates, then the descriptive statistics from the data in table 8.7 offer a summary of these varied dimensions.

TABLE 7: Descriptive statistics for A. E. Berriman's table.

	Inches	Meter
Mean	20.04	0.51
Median	19.61	0.50
Standard deviation	2.57	0.07
Range	7.76	0.20
Minimum	17.48	0.44
Maximum	25.24	0.64

Table 8.7 Descriptive statistics for A. E. Berriman's table

The estimates in Berriman's table for Greek and Roman cubits align reasonably well with the Egyptian short cubit suggesting an average of approximately eighteen inches. This dimension is about two inches shorter than the overall mean in table 8.7. The full range of values is about eight inches from 17.5 to twenty-five. The varied origins for these data and previous values suggest considering a family of cubits accumulated from many geographic areas over many different times rather than view these differences as suspects of one exact dimension. Such variants may not be simple differences, or differences around an exact unit, but rather a composite of dimensions accumulated over a long chronological period from many geographical locations that cannot be disentangled. These multiple dimensions suggest local applications rather than simply differences about a single standard, which frustrates greater accuracy.

A rounded value of 18" seems common for this period. The Hellenistic cubit appears in line with what has been identified as the short cubit. Standardization of the cubit began during Hellenism coinciding with Alexander's conquests in the Middle East. Its standardization was probably increased greatly under the Roman Empire from the influences of war, travel, and trade. These influences contributed to bringing the cubit into a more standard operational unit. Roman engineers in viaduct, bridge, and road construction brought standardization throughout the empire.

Cubits were employed through Antiquity to the Middle Ages and continue even today in some parts of the East. Continued usage prevailed for measuring textiles by the span of arms with subdivisions of the hand and cubit in less industrialized countries.

Moving forward to Da Vinci (1452–1519), we have his specifications and commentary on Vitruvius Pollio (first century BCE) for the human figure and its dimensions. They can be summarized as fractions of a six-foot man as given in table 8.8.

Unit	Inches
Finger	0.75
Palm	3
Foot	12
Cubit	18
Height	72
Pace	72

TABLE 8: Human dimensions relative to the six-foot male.

Table 8.8 Human dimensions relative to the six-foot male

Figure 8.2 gives the famous picture associated with these dimensions. The unit given shows one more example of the dimension of the cubit.

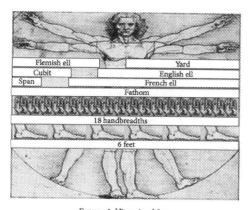

FIGURE 1: Vitruvian Man.

Figure 8.1 *Vitruvian Man*

The figure of the *Vitruvian Man* by Leonardo da Vinci depicts nine historical units of measurement: the yard, the span, the cubit, the Flemish ell, the English ell, the French ell, the fathom, the hand, and the foot. The units depicted are displayed with their historical ratios. In this figure, the cubit is 25 percent of the 6′ individual (about eighteen inches). We are reminded once more of the importance of the human figure for establishing units of measure.

Another example from this period comes from the *Autobiography* of Benvenuto Cellini (1500–1571). In describing his casting of Medusa, Cellini's narration uses cubit to illustrate length as casually as today we might use foot or yard. At least in this context, if not others, the cubit appears of common usage. How much more generalized a cubit dimension was through this time period is not known. By the time of the French

235

Revolution, the Committee of Weights and Measures had abandoned the cubit (among other dimensions) in favor of the metric system.

The Human Cubit

The history of metrology provides interesting data on the varied dimensions of the cubit. Metrology first utilized the human figure in establishing dimensions. History to this point suggests that a value of about 17–18" seems average and most common.

Sir Francis Galton (1822–1911) offers data gathered from the investigations he conducted. Galton deserves recognition as one of the first investigative anthropometrists. He was a scientist producing some of the first weather maps for recording changes in barometric pressure and strategies for categorizing fingerprints. Galton stands out for his investigations involving thousands of subjects. Some investigations were conducted at the International Health Exhibition in London held 1884–85 and at other field locations. Galton had earlier made an analysis of famous families from which he compiled *Hereditary Genius* and later in *Natural Inheritance*. He maintained a lifelong interest in determining the physical and mental characteristics of groups of individuals.

Not only did Galton collect data from his laboratory on human subjects, he investigated statistical techniques for analyzing tables, graphs, and plots of data. In doing so, he created the origins of what is now recognized as correlation and regression analysis. Correlation became more formally developed by Karl Pearson as the product moment correlation coefficient. It has become the most known and used statistical procedure of our time. Other statisticians, especially Sir Ronald Fisher and John Tukey, have criticized the correlation coefficient for its abuse arising from simplistic applications and dubious interpretations. Nevertheless, the correlation coefficient remains a popular analytic technique. Pearson also produced three volumes on the life, letters, and works of Galton.

Galton's data for the cubit of his day is given in table 8.9. It was taken from Stigler's *The History of Statistics*. Its original source is Galton, whose investigation gives data gathered from about 130 years ago on the forearm or cubit. Stigler indicated three of Galton's row totals were summed incorrectly. These sums were corrected in Table 8.9.

TABLE 9: Frequency of left cubit measure by inches.

Stature by inches	Under 16.5	Under 17	Under 17.5	Under 18	Under 18.5	Under 19	Under 19.5	Above 19.5	
71+	0	0	0	1	3	4	15	7	30
70	0	0	0	1	5	13	11	0	30
69	0	1	1	2	25	15	6	0	50
68	0	1	3	7	14	7	4	2	38
67	0	1	7	15	28	8	2	0	61
66	0	1	7	18	15	6	0	0	47
65	0	4	10	12	8	2	0	0	36
64	0	5	11	2	3	0	0	0	21
~64	9	12	10	3	1	0	0	0	35
Total	9	25	49	61	102	55	38	9	348
Inches	16.5	17	17.5	18	18.5	19	19.5	19.5	
Frequency	9	25	49	61	102	55	38	9	

Table 8.9 Frequency of left cubit measure by inches

Figure 8.2 summarizes the relative frequency of forearm/cubit lengths from Galton's data on 348 subjects given in table 8.9.

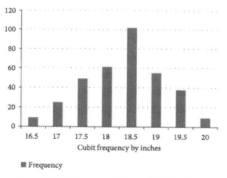

Figure 2: Cubit frequency by inches for 348 subjects.

Figure 8.2 Cubit frequency by inches for 348 subjects

Figure 8.2 indicates that the modal category of forearm/cubit measures for Galton's sample was 18.5 inches. The frequency distribution of forearm measurements is somewhat balanced. This might be expected given that these measures would be determined by chance through heredity. This was Galton's viewpoint and emphasis. Consequently, this data and other data moved his interest to eugenics. Many other English scientists and statisticians shared this interest: Fisher, Pearson, Haldane, Cattell, and others. Galton (and the others) received considerable criticism for taking this position. However, it was as a scientist and compiler of human data that led Galton to draw his inferences. His pronouncements concerning eugenics do not smack of a political or personal agenda. One may disagree, but it is important to understand that Galton's work was focused upon data and methodology as the basis for forming his conclusions.

The mean for the Galton sample of 348 persons in table 9 was almost eighteen inches, bringing estimates of a center location (i.e., mode, median, and mean) in sync with an approximate normal distribution as shown in table 10.

TABLE 10: Millimeters and inches of the left cubit.

	Millimeters	Inches
Mean	67.06609	17.83621
Standard error	0.126798	0.042699
Median	67	18
Mode	67	18
Standard deviation	2.365384	0.796541
Sample variance	5.595043	0.634478
Kurtosis	−0.9142	−0.42833
Skewness	−0.09243	−0.16653
Range	8	3.5
Minimum	63	16
Maximum	71	19.5
Sum	23339	6207
Count	348	348

Table 8.10 Millimeters and inches of the left cubit

From Galton's data summarized in figure 8.2 and tables 8.9 and 8.10, about 2 percent had forearms at 16.5" or less, and 2 percent had forearms greater than 19.5". Approximately 63 percent or 218 persons and close to two-thirds of the 348 person sample are within one half inch + or − the mean of 18.3 inches or almost 18.5" if rounded off. About 95 percent vary less than an inch above and below the mean estimate. Rounding from these frequencies makes these values approximate, but they still provide a generally useful summary from his sample. Skewness and kurtosis appear as minimal influences on the distribution further confirming a balanced distribution.

Figure 8.4 provides a three-dimensional view of Galton's data. It usefully shows the clustering of values along the center diagonal from the upper left to lower right. Galton's figures were not shown as three-dimensional, but he recorded the frequencies at each intersection of his two-way table, which were used to produce this three-dimensional figure. Pondering his data gave rise to Galton's work on association/correlation for which the word regression has now evolved being derived from his efforts to interpret what this and other data express. See Stigler for more details on Galton's analytic methods. These matters are not directly connected to the issues of cubit length and therefore not discussed here. However, the relationship of cubit to stature is useful and can be compared to Da Vinci's estimate.

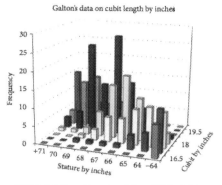

Galton's data on cubit length by inches

PIGURE 3: A three-dimensional view of Galton's data.

Figure 8.3 A three-dimensional view of Galton's data

Stigler indicated "Galton's ad hoc semigraphical approach gave the correlation value." This was Galton's approach prior to the Pearson product moment correlation. Figure 8.4 is a plot of data from Table 8.9 with a linear regression line showing the variation in forearm/cubit at each level of stature. It is very important to note the wide variation of left cubit measures (vertical) for each indication of stature (horizontal). Individual differences in the cubit/forearm are clearly evident at each point of stature thwarting anything more specific than a generalized indication for the forearm/cubit from Galton's data. The shared variance between stature and cubit is about 57 percent (r = .76) suggesting these two variables are related but not perfectly.

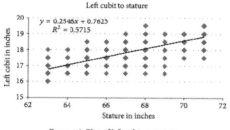

Left cubit to stature

FIGURE 4: Plot of left cubit to stature.

Figure 8.4 Plot of left cubit to stature

Several questions emanate from Galton's data regarding forearm length or the cubit: (1) How representative is this sample of the general population? (2) How much change, if any, in human dimensions has occurred from ancient times and over the hundred plus years from Galton's sample to the present day? (3) Are there gender differences or other sources of influence and (or) bias?

From what we know of Galton's methods, there appears no indication of outright bias. Stigler in chapters 8, 9, and 10 of his book raised no questions when describing Galton's data and methods for analyzing data. Galton's samples were large and often in the thousands. This cubit sample is moderate in scope. Galton was aware of gender differences and utilized 1.08 as a correction factor for male/female differences.

However, there is little information regarding sample representation. It appears that Galton was generally fastidious in his investigations. He utilized gatherings of the general population from which to procure his samples and make his measurements. Given that right-handedness predominated, Galton measured the left hand to avoid what might result from possible environmental influences upon the mostly dominant right hand. Volunteering could be a potential source of bias, but volunteering probably allowed a larger sample of individuals. He paid individuals a modest amount to participate, not unlike what is sometimes done today.

Johnson et al. (1985) reviewed and reanalyzed Galton's original data. They report on mean scores, correlations of the measures with age, correlations among measures, occupational differences in scores, and sibling correlations. A correlation of cubit/forearm to stature indicated the former was about 25–27 percent of stature. Nothing further is added to a knowledge of forearm/cubit dimension by their work.

Relevance of the forearm/cubit length in more recent times comes from anthropometric dimensions utilized in industrial psychology and applications to the clothing industry. Data from Mech (2010) gives more recent data on human dimensions, including the forearm. Forearm lengths reported for percentiles 5, 50, and 95 are given in Table 8.11.

TABLE 11: Forearm percentiles for an unidentified British population.

Percentile	5	50	95
Male	440	475	516
Female	400	430	460

Table 8.11 Forearm percentiles for an unidentified British population

These percentiles are from an unidentified British sample ages nineteen to sixty-five. Lacking more information, one can only compare and contrast these dimensions to previous samples discussed earlier. These males had a median cubit measure of 475 mm or ~18.7 inches. Females measured a slightly shorter median measure of 430 mm or ~16.9 inches. Mech indicated a median value close to that given in table 8.9 for Galton's data or ~18.7 to ~18.3.

The Lean Manufacturing Strategy reports a forearm mean = 18.9', standard deviation = 0.81', minimum = 15.4', and maximum = 22.1' based on data from McCormick. Nothing further is given regarding this sample and its characteristics.

There are numerous sites and organizations providing carefully determined dimensions for the human body. However, these dimensions are developed to serve the clothing industry and furniture design, adding nothing to our knowledge of the contemporary forearm/cubit dimension.

The anthropometry database ANSUR obtained from http://www.openlab.psu.edu/ gives a table of percentiles for the horizontal measure made "from the back of the elbow to the tip of the middle finger with the hand extended," that is, cubit. The sample was comprised of unidentified male army recruits.

The ANSUR data sample in table 8.12 provides descriptive statistics for the right male forearm plus extended hand in millimeters. The mean for this quite large contemporary sample is about one inch greater than the short cubit reported much earlier. So is the median although the mode is slightly less. The sample appears reasonably balanced, but the variation indicated by the standard error, standard deviation, and range show this human dimension to vary. Variation has been encountered before in the reporting of earlier samples.

TABLE 12: Elbow-fingertip length percentile distribution in millimeters.

(a)

1st	2.5th	5th	10th	25th	50th	75th	90th	95th	97.5th	99th
435	442	448	455	468	483	499	515	523	532	542

(b)

Mean	484.04	(19.05 inches)
Standard error	0.55	
Median	483	(19.01 inches)
Mode	472	(18.58 inches)
Standard deviation	23.32	
Sample variance	544.09	
Kurtosis	0.43	
Skewness	0.22	
Range	192	
Minimum	386	
Maximum	578	
Count	1774	

Table 8.12 Elbow-fingertip length percentile distribution in millimeters

Discussion

The varied dimensions for the historical cubit of ancient times and places speak to a variation in the dimension itself. Two major units predominate; one estimate centers around eighteen inches, and the other around twenty inches. There are other variations, some smaller and some much greater. There is too wide a geographical area and too great a chronological time period to consider any of these latter variations normative. Each variant was more likely to be locally relevant rather than widely prominent. Only in the Greek and Roman Empires did these values coalesce to somewhat of a standard through war, trade, and construction.

How has the human physique changed over time? Roche reported that rates of growth during childhood have increased considerably during the past fifty to a hundred years. He indicated increases in rates of growth and maturation for all developed nations but not in many other countries. There were recorded increases in length at birth in Italy and France but little change in the United States. An increase in childhood stature of 1.5 cm/decade was for twelve-year-old children. The increase in stature for youth was about 0.4 cm/decade in most developed countries. The changes in body proportions during recent decades were reported as less marked than those in body size. Leg length increased more than stature in men but not in women. Roche further indicated that changes in nutrition alone could not account for the trends that exceed the original socioeconomic differentials. In the United States, Roche reported there have been per capita increases in the intake of protein and fat from animal sources, decreases in carbohydrates and fat from vegetable sources, and some changes in caloric intake. It is not clear that these changes constitute better nutrition, stimulating growth. The trends could reflect environmental improvements, specifically changes in health practices and living conditions leading to improvements in mortality rates and life expectancy. Nutrition varies even in developed countries. Roche reported genetic factors play a small role in causing trends. However, the data speaks to considerable variation among contemporary samples as also noted in Galton's data.

Overall, it seems unwise to be overly fastidious about any contemporary value for the cubit when such samples are vaguely described. For any comparison of contemporary dimensions reported, there are few characteristics given by which to judge sample representation. The contemporary estimates appear somewhat close together and suggest at least for these samples no great change has occurred over the years, but lacking valid data, we cannot be sure. Without more sample definition, any fastidious analysis appears unwarranted. The Galton values are likely to have been local and relevant to a British sample. Nowadays, samples are more likely to reflect the role of immigration with whatever additional effects this might bring to bear on determining national human dimensions. In general, Europeans are taller than Asian/Middle East peoples and Americans are taller than Europeans. These are generalizations from gross estimates. Komlos and Baten (1998) have made a comprehensive analysis of stature over centuries. The striking feature of their tables is the variation in values across time periods. Individual variation was also observed in Galton's data. However, systematic sampling and sample details must accompany any data before estimates can be more than gross general indications.

A variety of circumstances address the cubit, but most of them offer little specific information beyond what has already been presented. These biased sites typically serve some agenda, often religious or personal. Overall, even these sites typically report the two major dimensions for the cubit at eighteen inches or twenty inches.

The cubit as a dimension remains useful. We take the cubit (hand and foot) wherever we travel. Knowing personal dimensions can sometimes prove useful for making quick albeit gross estimates. The 18" ruler is a very handy device whenever measures just beyond a foot ruler are required, especially when it is necessary to draw straight lines for a length just beyond twelve inches. Tape measures are a boon but not for drawing lines.

It appears that we might content ourselves with a cubit length of eighteen inches as a somewhat consistent dimension for the cubit. Even as the foot evolved from a specific albeit arbitrary personage, any assemblage of them leads to an abstract dimension, so the cubit could justify more application as a 0.5 yard and (or) a 0.5 meter. Further prominence of either or both of these units might prove more useful than first surmised.

Box 25: Thales and Rasch

Teaching the geometry of the right triangle is frequently enhanced by the story of Thales (c. 640–550 BCE), who applied the property of similar triangles to measure the height of a pyramid. He deduced that the shadow of the pyramid (QB) and, simultaneously, that of a vertical stick (AB) placed at the end of the shadow of the pyramid produce the following property of similar triangles: the height of the pyramid (PQ) is to the height of the stick (AB) as the length of the shadow of the pyramid (QB) is to the length of the shadow of the stick (BC) (i.e., PQ/AB = QB/BC). Two comparisons to Rasch are evident in this illustration:

1. The practical solution to finding the height of any object is determined by an abstract principle, not by data. Rasch always sought to find general solutions to measurement problems, not merely to produce a description of data (Stone and Stenner 2014, 2016). Similarly, we follow Newton's theory of gravitation, not his data.
2. The solution to the problem of determining the height of any object is independent of the height of the stick utilized and independent of the height of the object that is to be measured.

Thus, we see that parameter separation and general objectivity are very old and very useful ideas.

Box 26: Time

Augustine (1957) writes in chapter XIV of his *Confessions*:
Neither time past nor future, but the present only, really is.

> What, then, is time? If no one ask of me, I know; if I wish to explain to him who asks, I know not. Yet I say with confidence, that I know that if nothing passed away, there would not be past time; and if nothing were coming, there would not be future time; and if nothing were, there would not be present time. Those two times, therefore, past and future, how are they, when even the past now is not; and the future is not as yet? But should the present be always present, and should it not pass into time past, time truly it could not be, but eternity. If, then, time present—if it be time—only comes into existence because it passes into time past, how do we say that even this is, whose cause of being is that it shall not be—namely, so that we cannot truly say that time is, unless because it tends not to be? (224)

Do these words indicate careful reflection, insight, or casuistry? Past is not now and future is not yet, says Augustine. Only the present exists emerging from the future and entering the past. Such definitions or declarations remain speculative and philosophical.

Wittgenstein (1958) referred to Augustine's words in Aphorisms 89, 90, 607, 608 found in *Philosophical Investigations*. Wittgenstein conjectured philosophy to be a correction of thinking (i.e., grammatical, hence thought, correction). This approach speaks to understanding an intangible concept—time, for example, by making (fabricating, constructing, engineering) the intangible to become tangible in some manner. Therefore, Wittgenstein advocates the only way possible to understand intangible terms or concepts is to fabricate them.

Earlier, in *Tractatus Logico-Philosophicus* (1922), Wittgenstein wrote, "We cannot think of temporal objects apart from time. "Space, time and color are forms of objects" (# 2.0251). Time does not appear as an "atomic fact" for Wittgenstein but always in conjunction with something—"that the sun will rise tomorrow is an hypothesis" (# 6.36311).

Newton (1947) from the General Scholium attached to *Principia* comments:

> Absolute, true and mathematical is a time, of itself, and from its own nature, flows equitably without relations to anything external, and by another name is called duration: relative, apparent, and common time, is some sensible and external (whether accurate or equitable) measure of duration by means of motion, which is commonly used instead of true time such as an hour, a day, a month, a year. (6)

Newton argues that time flows. By another name, duration is determined by some defining process. Hence, an hour, a day, a month, or a year, time can be determined by motion; initially sense-driven observations but subsequently expressed by some devised measure.

Early inventions for tracking time: an origin, a unit, and its precision utilized anthropomorphic, terrestrial, celestial, epic, personage, and eventually committee standards.

1. Anthropomorphic: human palm, span, cubit, pace
2. Celestial: revolution of earth—day and night; revolution around the sun—year, revolution of selected star around earth—sidereal day.
3. Terrestrial: sundial, water clock, marked candle
4. Epic: founding of Rome, Christianity, Islam
5. Personage: Julian calendar, Gregorian calendar, a king's royal foot, and so on
6. Committee: metric system, the second, one of seven SI basic standards

A unit and origin are determined arbitrarily by virtue of some characteristic determined from the characteristics of the identified object or process. Determination by some consensus of strategy and instrumentation is generally regarded as more scientific and a move toward universal application. The second is an example.

The Second
The second is defined as:

> The time that elapses during 9,192,631,770 (9.192631770 x 10^9) cycles of the radiation produced between the two hyperfine levels of the ground state of the cesium 133 atom.

Once defined as 1/86 400 of the mean solar day, the "mean solar day" was inaccurately producing irregularities because of the irregular rotation of the Earth (BIPM 2013).
Even atomic clocks lack precision.

> A leap second is the one-second adjustment occasionally applied to Coordinated Universal Time (UTC) in order to keep its time of day close to the mean solar time, or UT1. Without such a correction, time reckoned by Earth's rotation drifts away from atomic time because of irregularities in the Earth's rate of rotation. Since this system of correction was implemented in 1972, 27 leap seconds have been inserted, the most recent on December 31, 2016 at 23:59:60 UTC,https://en.wikipedia.org/wiki/ Leap second - cite note-Bulletin C 52-1 the next leap second has currently not been announced, but is projected to be after 2018. (BIPM 2013)

The Measuring Mechanism

A measurement mechanism is the name Stenner, Smith, and Burdick (1983) give to those manipulable features of the instrument that cause invariant measurement outcomes for objects of measurement that possess identical measures. A measurement mechanism explains by opening the black box and showing the cogs and wheels of the instrument's internal machinery. A measurement mechanism provides a continuous and contiguous chain of causal links between the encounter of the object of measurement and instrument and the resulting measurement outcome (Elster 1989). We say that the measurement outcome (e.g., raw score) is explained by explicating the mechanism by which the measurement outcome is obtained:

1. A substantive theory composed of detailed specifications of mechanisms resulting in more intelligible explanations.
2. A focus on mechanisms reduces theoretical fragmentation by encouraging consideration of the possibility that many seemingly distinct instruments (e.g., thermometers, clocks, reading tests) with different item types and construct labels may in fact share a common measurement mechanism.
3. The requirement for mechanismic explanations helps to eliminate spurious causal accounts of how instruments work.

Measurement mechanisms as theoretical claims make point predictions under intervention (Woodward 1997, 2003). When we change (via manipulation or intervention) either the object measure (e.g., reader experiences growth over a year) or measurement mechanism (e.g., increase text complexity by 200L) the *mechanismic* narrative and associated equations enable a point prediction on the consequent change in the measurement outcome (i.e., count correct).

Causal models (assuming they are valid) are much more informative than probability models: A joint distribution tells us how probable events are and how probabilities would change with subsequent observations, but a causal model also tells us how these probabilities would change as a result of external interventions … Such changes cannot be deduced from a joint distribution, even if fully specified (Pearl [2000] 2009, 22). A mechanismic narrative provides a satisfying answer to the question of how an instrument works.

Herbert Simon (1969) distinguished between two strategies for conducting science. He designated the first *natural science,* which is "knowledge about natural objects and phenomena," and the second *artificial science* (i.e., fabricated), "meaning man-made as opposed to natural" (4). His terminology advocates words such as producing, manufacturing, and constructing to describe the processes for inventing artificial science, and he writes,

> It has been the task of the science disciplines to teach about natural things: how they are and how they work. It has been the task of engineering schools to teach about

artificial things: how to make artifacts that have desired properties and how to design. (5–6)

The boundaries for sciences of the artificial:

1. Artificial things are synthesized (although not always or usually with full forethought) by man.
2. Artificial things may imitate appearances in natural things while lacking, in one or many respects, the reality of the latter.
3. Artificial things can be characterized in terms of functions, goals, adaptation.
4. Artificial things are often discussed, particularly when they are being designed, in terms of imperatives as well as descriptions (p. 5-6).

The social sciences are in the domain of artificial science. We see measuring mechanisms as "engineering" an artifact constructed from substantive theory, instrumentation, and supportive data. The history of length, temperature, and time illustrate the developmental role of engineering, construction, and fabrication to produce measures.

Stenner, Stone, and Burdick (2009) have addressed the fundamental difference between constructing measuring mechanisms for producing measures and using the Rasch model to describe data. The major difference is the emphasis of the former on "constructing" and the importance of substantive theory driving the investigation. Temperature is the ubiquitous example for mechanisms for measuring constructed by analogy (temperature analogous to a column of mercury) together with a long history of instrument development. Chang ([2004] 2007) has provided a comprehensive history for what he labels *the invention of temperature*, which sparks our inquiry into investigating time. The continuous development of theory, instrumentation and data produces the developmental and continuous "inventions" that have given us more precise and more ubiquitous tools for measuring temperature and for measuring time/duration.

The "invention of time" reveals a similar process for the construction of mechanisms from observation of the heavens and the observational "measures" invented to the most sophisticated devices for recording time. The key constructive processes for constructing measures are the following:

1. Comparison by which similarities and differences are determined

2. Order by which elements are sequenced
3. Spacing from which a unit is derived
4. Origin for a point referenced

Temperature for most of us is the sensation of heat or cold we experience in our environment. A typical thermometer commonly uses an expansion tube of mercury to accomplish the task of graduating the level of mercury and determining an associated measure. Water and alcohol among other elements were investigated in arriving at the choice for mercury for many household thermometers. Variations abound on the way to producing utility in measures. Celsius ranges between freezing water at 0°C to boiling at 100°C, while Fahrenheit ranges between 32°F for freezing water and 212°F for boiling. There are numerous other scales. The theoretical Kelvin K dovetails with the Celsius scale. Freezing water and boiling water establish a two-point common standard. Although the matter of temperature is complex, a common household thermometer satisfies utility for most occasions.

A complex variable has been reduced to a simple one. Rasch ([1960] 1980) in discussing models in classical physics remarked:

> Nonetheless it should not be overlooked that the laws do not at all give an accurate picture of nature. They are *simplified descriptions of a very complicated reality.* (10, our emphasis)

This point seems rarely appreciated to judge from the voluminous amount of commentary in the social sciences citing how "complicated" reality is and how difficult it is to model. Physics has progressed admirably formulating "simple" laws to model complex matters. Scientists appreciate complexity, but nature cannot be understood when complexity is made a stumbling block to understanding. In such instances, emphasizing complexity obfuscates understanding and knowledge. The history of temperature and time reminds us that complexity can be modeled in a simple fashion if only we can find a useful way to do so. Bridges and roads are not engineered to last forever. All variables constructed to be utilitarian must be constantly monitored and maintained.

Time or duration likewise shows a progression from sensory observations to constructed units, strategies, and instruments that provide a means for measuring time. The implications of inventing temperature and inventing time can be important for understanding the essence of measuring. We measure by analogy. For measuring time, we have had sand trickling through an hourglass to moving hands on a clock face. No matter how sophisticated, the device (cesium clock) analogy prevails in some form, and precision must be monitored.

A validly constructed instrument must emanate from a single, unified variable. The problem is to devise and construct one. There is only one construct variable but many ways to divide and express it. Validity rests on achieving instrument integrity and invariance. Everything else is peripheral to this problem and only serves to confuse the matter. General application and utility constitute validity with unique exceptions. Utility is an important aspect of measuring. The choice between two explanations, complex versus simple (Occam's razor), favors the simple as the useful one. Utility implies understanding.

Sensation to Gradation

The continuous cycle of theorizing, investigating, and experimenting is the history of science. The strategies applied to "inventing temperature" (Chang [2004] 2007) or inventing time are models for the construction of other variables. The story of inventing temperature and inventing time is a record of the developmental history beginning with a modestly differentiated sensation of hot/cold or long to short duration to a multiplicity of measures constructed by a continuous refinement of substantive theory, technological advances in instrumentation, and the confirmation of data substantiating predictions. The course for developing any variable can follow in a similar manner.

Establish an origin such as a count of astronomical units (AU = average distance between the earth and the sun), or a two-point standard as between freezing water and boiling water. This unit, large or small, may be divided or expanded as deemed useful. Convert the count to a measure using Rasch's equation.

Box 27: Constructing Measures

Herbert Simon (1969) distinguished between two strategies for conducting science. He designated the first *natural science*, meaning "knowledge about natural objects and phenomena," and the second *artificial science*, "meaning man-made as opposed to natural" (4).

His terminology advocates using words such as producing,

manufacturing, and constructing to describe the processes for artificial science:

> It has been the task of the science disciplines to teach about natural things: how they are and how they work. It has been the task of engineering schools to teach about artificial things: how to make artifacts that have desired properties and how to design (p. 5-6).

> The boundaries for sciences of the artificial:

> 1. Artificial things are synthesized (although not always or usually with full forethought) by man.

> 2. Artificial things may imitate appearances in natural things while lacking, in one or many respects, the reality of the latter.

> 3. Artificial things can be characterized in terms of functions, goals, adaptation.

> 4. Artificial things are often discussed, particularly when they are being designed, in terms of imperatives as well as descriptives. (5–6)

The social sciences are in the domain of artificial science. Measuring mechanisms are engineered and artifacts constructed from substantive theory, instrumentation, and supportive data. The history of length, time, and temperature illustrate the role of engineering, construction, and fabrication in the production of measures.

Stenner, Stone, and Burdick (2009) have addressed the fundamental difference between constructing measuring mechanisms for producing measures and models that only describe data. The major difference is the emphasis of the former on "constructing" and the importance of substantive theory driving the investigation.

In the Rasch model, there is a triad consisting of an outcome resulting from the interaction of objects and agents, or items and persons in psychometrics.

· Outcome (Reaction)

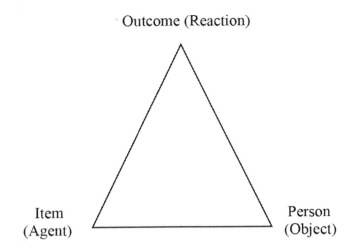

Item   Person
(Agent)   (Object)

Rasch's (1977) specification of "outcome" or "reaction" is underlined in the following quotations:

> Every object O of C may enter into a well-defined contact C with every agent A of A and every such contact has an 'outcome' R. (75)

> The term 'objectivity' refers to the fact that the result of any comparison of two objects is independent of the choice of the agent A within A and also of the other elements in the collection of objects O; in other words: *independent of everything else within the frame of reference except the two objects which are to be compared and their observed reactions (p. 77).*

> Main Theorem I. Let objects and agents in the bifactorial determinate frame of reference F be characterizable by scalar parameters ω and α, and reactions by a scalar reaction function of 'convenient' mathematical properties

> $$\xi = \varrho\,(\omega, \alpha). \ (79)$$

> Main theorem I deals with reactions produced through contacts between two kinds of elements denoted by objects and agents .... (81)

Main theorem I states that specifically objective comparability of pigs as well as of age levels in a scalar parameterization with respect to growth requires the reaction parameter, i.e.,

$$\upsilon_{t+1} - \upsilon_t \text{ to be latently additive } \ldots. (90)$$

These references and more apply to an outcome/reaction from a conditional model in which one element [agent or object] is held constant [canceled/eliminated] while two of the other [agent or object] elements are compared. Simple triads of this type are common in mathematics and physics, e.g. $F = MA$, $E = MC2$. The essential point is how the equation $D = RT$ permits one to solve for any of the three variables, distance, time, or rate, by using any two variables to produce the third.

Evaluating the consequences of any substantive theory employing the Rasch Model is given below:

Meta-Model for evaluating the Rasch Model RM and Substantive Theory ST

Rasch Model (RM)

	Fit	Misfit	
Fit	1	3	Substantive
Misfit	2	4	Theory (ST)

Cell	Column and Row	Conditions to determine
1	RM fit ST fit	Meets doubly prescriptive model, validate
2	RM fit ST misfit	Lack of correspondence between RM and ST, reexamine ST
3	RM misfit ST fit	Lack of correspondence between RM and ST, reexamine RM
4	RM misfit ST misfit	Examine what is causing problems in both RM and ST

Under quality control conditions, both aspects should be examined and reexamined.

The *doubly prescriptive* approach also requires emphasizing the fabrication/manufacturing process of data. We do not "find" data but must produce data appropriate to the investigation. This will require reorienting investigators to more appropriate terminology and mind-set regarding data.

"Recipe" may seem a strange and inappropriate term not suitable to science, but it does fit the Simon fabricating model for artificial science and nicely illustrates how we create and follow designs. We can create a recipe to describe what we intend to do (i.e., a prescriptive model), and we follow a successful recipe to assure that successive replications will occur Whether or not you like McDonald's food, the one thing you must appreciate is that you cannot separate their products by location. Every product is engineered to the same specifications regardless of the location of the restaurant. This is true for any manufacturing program that follows a clear design (recipe) for producing a product. It is the secret of success in fabricating a process.

A recipe assures an idea, theory, sample observations, instrumentation, and so on are the prescriptive elements of a substantive theory. Without the recipe, the results cannot be examined to determine how the constituent pieces work together, which ones are working as specified, and which need attention in order to improve the experiment. A valid recipe results in a specification equation that gives the specifications for manufacturing an outcome.

While "recipe" sounds informal, the production process is similar whether one speaks of cooking, manufacturing, or any other form of fabrication. To be more explicit, we strive to produce a mathematical equation, chemical formula, logistical plan, fabrication schedule, etc. The emphasis of all of these operations is upon production with quality control over the fabrication process.

Such a consideration suggests the following *Ladder of Quality Control* for evaluating a *Doubly Prescriptive Model*:

Ladder of Quality Control

RM	ST	Result
+	+	Specification equation
+	-	Index only
-	+	Idea
-	-	Nothing

Here we have prescribed Rasch Model RM and Substantive Theory ST as the two constituent elements of the *measuring mechanism*. The RM equation presupposes the raw scores to be linearized so they must be ordered inasmuch as the equation cannot make sense out of disarray in data. Misfit analysis is the strategy that identifies disorder in the RM. ST must also imply order if any sense is to be made of observations. Manifestations of depression, as in any example, must be ordered by some criterion such as severity of illness if the variable observations are to make sense.

The Doubly Prescriptive approach requires the fabrication, construction, manufacturing process for data. We do not "find" data but must produce data appropriate to the investigation. Stone (RMT 1996, 517) in *Data: Collecting or Manufacturing* indicates, "Manufacturing data describes the process we require. We do not 'find' data by beachcombing ... We produce data following a theory with an intention ... Quality control over production is absolutely necessary."

The social sciences can learn much from this simple exposition. Determining whether or not a variable has quantitative properties is not required. The "causal process" is the working paradigm. Practicing science requires more than an amalgamation of isolated "facts." A measuring mechanism is required with a *model/theory* by which to evaluate data. Rasch (1961) addressed this problem with a theory, a model, and specific data examples. His goal was "replacing qualitative observations by quantitative parameters" (331).

We call attention to three instruments that utilize the "height of a column of mercury." While further modification in their development has occurred, we choose to focus upon the commonality they share. The height/length of a tube of mercury is analogous to its respective attribute—human/environmental cold/heat; air pressure for inferring atmospheric changes; and blood pressure (i.e., systolic and diastolic, high/low indications of pulse). Each one may show a change in mercury height associated to a change in a human or environmental condition leading to these results:

Thermometer mercury height—human/environmental temperature
Barometer mercury height—atmospheric conditions
Sphygmomanometer mercury height—arterial pulse pressure

From each of these *measuring mechanisms* (Stenner, Stone, and Burdick 2009), important inferences may be drawn:

Measuring Mechanism Inference/Application

------------------------------ ----------------------------

Thermometer	—human well-being, fever or lack thereof, health status
	—status of environmental heat and cold, possible weather effects
Barometer	—air pressure and possible changes in weather
Sphygmomanometer	—cardiac/arterial condition, health status

Each of the three measuring mechanisms utilizes millimeters of mercury in a sealed tube as the units for providing linear measures. Not only is there an analogy between millimeters of mercury and these resultant states, the connection is algebraic and symmetric. Knowing one side of the equation, we can make predictions to the other side. Sensing a fever, we might use a thermometer to determine if the elevation is critical. A drop in the barometer might precede a change in weather, while in the midst of a hurricane, a low barometric pressure would be predicted. A sustained high blood pressure might cause a physician to introduce treatment, while the onset of death could be indicated by a continuing decrease of pulse. We infer these types of consequences and predictions daily from interplay between both sides of the equation.

The following diagram illustrates the information discussed:

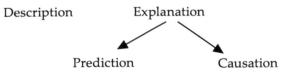

Description   Explanation

Prediction   Causation

Theory/Equation Physical data support "over wide range of changes"

Box 28: Causal Rasch Model versus Descriptive Rasch Model

Earlier commentary (Stenner, Stone, and Burdick 2011) introduced the term causal Rasch model (CRM) requiring a substantive theory to be joined to a set of predictive statements. A constrained Rasch Model provides a stronger context for interpreting measurement outcomes, especially residuals and anomalies. A formative approach in contrast to a reflective one (Stenner, Burdick, and Stone 2008) distinguishes predictive power from association for which the latter may be only temporal and data bound.

This difference—between prediction via a causal Rasch model (CRM) and association via a descriptive Rasch model (DRM)—assists in appreciating power by manipulation rather than solely by description. The latter may have predictive expectations, but it is not conceptualized or demonstrated sui generis.

These two approaches, CRM and DRM, suggest similarity to a purported contrast between external and internal validity (Campbell and Stanley 1963; Cook and Campbell 1979; Shadish, Cook, and Campbell 2001). Furthermore, description resting solely upon association suffers, as do factor analysis and regression analysis, when model specification and hypotheses are lacking.

Causal Rasch models (CRM) propose a theory/data relationship utilizing manipulation that can be evaluated by its predicted outcome. Descriptive Rasch models (DRM) absent a specified goal may be suggestive, but they are not strong advocates for understanding sans theoretical underpinning. We suggest it is necessary to distinguish *description* from *prescription* in our thinking and practice. The former, however carefully comported, remains solely a fixed examination of data. Prescription contains a formulary of explanation via prediction and manipulation. While validation remains a critical aspect for each approach, any prediction is always the function of a clearly focused goal and outcome. Stenner and Horabin (1992) suggested three stages:

Stage 1. Instrument calibration based on personal knowledge, intuition, and subjective analysis
Stage 2. Data-based instrument calibration
Stage 3. Theory-based instrument calibration

We may further determine three ordered levels within each state—descriptive and prescriptive.
Descriptive:

1. I wonder what this means.
2. Here is what we see.
3. This [perhaps] is what it means.

Prescriptive:

1. Adopted as if (Hans Vaihinger) true (e.g., operational explanation)

2. Abduction (C. S. Peirce) a tendered working hypothesis
3. A substantiated specification equation (theory plus data)

Verification and quality control over the process is essential. Consider a Winsteps map of the variable obtained from data: it may be schematized by agents and objects, but it remains descriptive. From an adopted operational definition springs abduction. C. S. Peirce (1998) writes:

> The mind seeks to bring the facts, as modified by the new discovery, into order; that is, to form a general conception embracing them. In some cases, it does this by an act of *generalization*. In other cases, no new law is suggested, but only a peculiar state of facts that will "explain" the surprising phenomenon; and a law already known is recognized as applicable to the suggested hypothesis, so that the phenomenon, under that assumption, would not be surprising, but quite likely, or even would be a necessary result. This synthesis suggesting a new conception or hypothesis, is the Abduction. (287)

Only when this process results from a substantive specification equation does is merit the label "prescriptive" in keeping with Rasch's ([1960] 1980) ideas as supported by these remarks:

> If a relationship between two or more statistical variables is to be considered really important, as more that an ad hoc description of a very limited set of data – if a more or less general interdependence may be considered in force – the relationship should be found in several sets of data which differ materially in some relevant aspects. (9)

> Once a law has been established within a certain field then the law itself may serve as a tool for deciding whether or not added stimuli and/or objects belong to the original group. (124)

The elements of risk and induction inherent to the process of prediction explain why many remain within the safety net of description, or mistakenly attribute description for prediction. Progress hinges upon making useful predictions and mobilizing thought applied to observations.

Box 29: KCT and Empirical Demonstration

Data and empirical verification are demonstrated only as the consequence of a tenable theory, not the other way around. Data from the development of "Knox's Cube Test—Revised" (Stone 2002) supports this position.

Knox's Cube Test (KCT) was first developed by Howard Knox (1914) as one instrument of a nonverbal battery developed to identify possible cognitive deficiencies in immigrants arriving at Ellis Island. Several revisions have subsequently been made, with all the versions summarized by Stone (2002b). Stone utilized the Rasch model to analyze the psychometric properties of the revised KCT whereby all the item difficulties and all the person abilities from these versions can be located on the same metric using rescaled logits (the logarithm of the odds).

The record page of the KCT-R report form indicates the items, score, mastery measure, performance criteria, and age norms. The record page is a "map" of the variable that defines its compass. Like any useful map, it identifies the locale defined by the item criteria—in this case the number of taps, the number of reverses, and the total distance. The goal is to ascertain the difference between the location of a person at the origin to what might be probable if the ability were greater. These criteria constitute the specification equation for the KCT-R.

The psychometric properties of KCT-R show exceptional stability for the items (Stone and Yumoto 2002). The KCT-R items exhibit unidimensionality with no differential item functioning identified.

New KCT-R items were developed using the three essential elements of the KCT-R items—taps, reverses, and distance. Multiple regression analysis was used to determine a linear equation for the KCT-R items. Rasch scaling and use of the multiple regression equation produced a linear transformation of the item parameter making compatible the result that should be obtained if the KCT-R holds desirable statistical properties. The Rasch scaled item difficulty (in logits) was selected as the dependent variable, with the number of taps, reverses, and distance as the three independent variables. The contribution of each independent variable was computed from regression analysis. The p-value matrix was analyzed using ANOVA to estimate the difficulty of each item and the values reexpressed as standardized Z scores. Finally, p-values were transformed into logits from the responses.

It is important to recognize that these new items were generated from an algorithm that describes the constituent elements of each item. Consequently, an item can be generated at any level of difficulty or with any combination of the characteristics of a KCT-R item, with no need for an item bank. There is no need to "bank" the items because they can be manufactured as required using the specifications equation. Security of the bank is never a problem because no one would ever seek to remember any of the thousands of combinations that can be generated. Determining the latent characteristics of an item makes the resulting algorithm the key to manufacturing items to order.

A residual analysis was made using Rasch fit statistics, principally the outfit and infit mean square of the items. These fit statistics provide essential information about item stability and any response bias that might be found from analysis of the residuals.

The multiple regression equation produced an R of 0.992 and R^2 of 0.984. The resulting equation was:

$$D_i = -10.42 + 1.766(\text{Taps}) + 0.004(\text{Reverses}) + 0.234(\text{Distance})$$

A plot of the two sets of KCT-R items with the observed item difficulty on the x-axis and the predicted item difficulty on the y-axis showed the items are located well within the 95 percent confidence interval. Also indicated are the correlations to earlier editions of the test. This results in a synthesis of more than 90 years of KCT versions published by eight different investigators. Again, nothing is so practical as a good theory.

These results illustrate the benefits that accrue when the research synthesis lens is focused on the instrumentation armament for a single construct. There are already too many nonexchangeable metrics in use for measuring human science constructs. We have three choices: (1) ignore the corrosive effects of interference produced from a variety of metrics, (2) invent temporary solutions, or (3) construct common metrics whenever and wherever possible. The choice appears to be obvious.

A measurement instrument embodies a construct theory: a story we tell about what it means to move up and down a scale (Stenner, Smith, and Burdick 1983). That theory should be submitted to rigorous testing to establish its validity. Unfortunately, given the large number of approaches to testing a construct theory, an egalitarian malaise has fallen upon psychometric practice, and all evidence for or against a construct theory is given equal weight. All evidence and, correspondingly, all methods are not created equal. In fact, one kind of evidence is preeminent—an explanation in the form of an equation that can explain the variation in item/ensemble difficulties or "endorsabilities." If we can't explain why one item is harder than another or is endorsed less often than another, we have only a very primitive understanding of what the instrument measures.

9

RASCH'S GROWTH MODEL

All of life is a trajectory.
—Steven Rose, biologist, WWU

Observations on growth have been constant throughout history. The pre-Socratics Plato and Aristotle made early observations. Shakespeare addressed seven stages with a soliloquy in *Hamlet*. Robbins, Brody, Jackson, and Greene (1928) and Courtis (1932) investigated growth models. Shock (1951) and Karkach (2006) are more recent examples from the many that might be cited.

The study of growth requires attention to several components: (1) investigating change over time in some dimension, (2) determining a functional relationship between these two magnitudes, and (3) deriving the resulting pattern, change, law, or generality. Components 1 and 2 are relatively straightforward tasks. Galileo rolled balls down an incline plan. His straightforward experiments yielded a time (t) to distance (y), which produced time (t) to velocity (v) whereby $v = 2t$ describes a "growth" function on speed.

The resulting outcome of any human growth study (component 3) depends upon how growth is to be expressed, and whether the growth is, for example, linear for height or length of the tibia, area for the spread of organisms, or mass (volume) for tissue and organs. Abstract variables for determining the growth of intelligence, ability, and achievement present more complex considerations (Andrich and Styles 1994; Dawson-Tunik Commons, Wilson, and Fischer 2005; McArdle, Grimm, Hamagami, Bowles, and Meredeth 2009; Wilson, Zheng, and McGuire 2012).

Williams cites, Georg Rasch pursued a life-long interest in growth models. He is less known for growth investigations, having been more popularly associated with psychometrics, particularly with item analysis (Rasch 1960). This diverted association is unfortunate inasmuch as Rasch devoted great time and attention to the growth studies applied to animals and persons together with econometric investigations. Fortunately, there is informal information provided by Andrich, Jensen, and Stene referenced in Olsen (2002), especially chapter 2 of her dissertation *Georg Rasch's Growth Model*. Her dissertation can be located at Rasch.org. Mention should also be made about what Olsen designates as Rasch's *Calcutta Notes* and *Danish Notes*. These substantial pages are not readily available.

We address Rasch's (1977) piglet study, which focused on evaluating the effects of different food supplements on the growth of piglets:

> ... an experiment with pig feeding [Source: Ludvigsen and Thorbek in Rasch (1977)] ... 14 pigs bought at birth ... taken from mothers age 20 days, experimental period, lasting age of 64 days ... fed same synthetic food ... different supplements to each of three experimental groups K-I, K-II, and K-III. Each day all pigs offered the same ration, [each] ate according to need, and digestion, combustion, personal processes within the chemical and physiological laws governing them ... the growth ... by weighing about every fifth day of the experimental ... period in Table 2. (87)

Table 9.1 data is taken from Rasch's Table 2 paper (1977).

Live weight by treatment code

Days	KI,24	KI, 60	KII,6	KII,11	KIII,24	KIII,21	Average
20	4.0	4.8	4.5	5.1	5.0	4.8	4.70
25	4.9	6.3	6.1	6.6	6.7	6.6	6.20
30	6.2	7.4	7.9	8.1	8.5	8.1	7.70
34	7.5	8.8	9.2	9.6	9.6	9.1	8.96
38	8.9	10.0	10.5	10.4	11.0	10.6	10.23
42	10.2	12.1	12.5	12.1	12.8	12.1	11.96
48	12.7	15.1	14.9	14.7	15.3	15.2	14.65
53	15.1	16.8	17.6	17.6	18.6	17.4	17.18
59	16.8	19.1	19.7	19.2	19.9	18.1	18.80
64	16.6	20.9	22.0	22.4	23.1	22.0	21.66
Average	10.3	12.1	12.5	12.6	13.1	12.4	12.2

Table 9.1: *Growth of 6 piglets from three different experimental series* Age Live weights: V_{it} (kg.)

Rasch made data transformations via logarithms together with growth plots:

Analyses after transforming the weight V_{it} (i: pig no., t: age minus 20 days) to their logarithms:

$$v_{it} = \log V_{it} \qquad\qquad \text{Rasch,} \qquad\qquad \text{VIII:2}$$

> ... to determine *an age parameter g_t, called the growth mode*, common for all pigs, such that *v_{it} plotted against g_t shows almost perfect linearity for each pig, which is then fully characterized by two individual parameters: location a_i and the slope b_i. The latter is called the individual growth rate:*

$$\log V_{it} = v_{it} = \alpha_i + \beta_i\,\tau_i\,. \qquad \text{Rasch,} \qquad \text{VIII:3 (p. 87)}$$

Equation VIII:3 is a structural equation explaining growth rate. Structural coefficients α and β specify the growth rate parameters. Models for complex behavior often utilize multiple structural equations, but VIII:3 models how the dependent variable weight v_{it} is influenced over time and feeding protocols by

$$\beta_i\,\tau_i + \alpha_i\,.$$

Rao (1958) described a variety of approaches to comparing growth curves. In doing so he gives $b = \sum y_i\,g_i / \sum g_i^2$ for rate of growth where y_i is the increase corresponding to time period g_i.

Rao (1958) indicated:

> ... this method of estimation was proposed by G. Rasch of Denmark during a course of lectures on growth curves he gave in India in 1951. This case can be shown to be the least squares estimate of the time metameter corresponding to the observed values under the assumption that the y_i are uncorrelated and have the same variance. (3)

Later in the paper, Rao explicates another example regarding weight:

> ... for the mean values of weights (absolute and logarithmic) for boys and girls at intervals of twenty days from birth, the time metameter with respect to which the growth is

expected to have a linear trend (which is estimated by the method of Rasch by averaging the weights of all the 27 children), and also values for gains in weight during the successive intervals. (14)

Measuring age in a particular way is important for Rasch, as Olsen (2002) indicates:

Rasch's question was whether similar types of organisms would grow in similar ways (the Rasch Calcutta notes, p. 26): *Now we may hazard the question whether the way in which a time interval counts at different ages is something specific to the type of organism considered. If so, we should by measuring the age in a particular way, get a uniform description of the growth curves for all organisms of the type considered.*

That is, for all organisms the τ_v function must be the same. Or equivalently, the deterministic model

$$\log (y(t)) = \alpha_v + \beta_v \, \tau(t) \qquad (2.3)$$

must apply to all organisms $v = 1, \ldots, n$. For further reference, we shall refer to (2.3) as *The Growth Model*. According to Andrich, Rasch referred to τ as *the metameter* in the 1970s (Int. Andrich, 06.02.2002). It was also given above in the Rao quote. Since this was the name Rasch specified, we shall use it (61).

We need to establish what growth implies by means of a prescribed outcome, equation, and application. Only then can we determine how individuals progress; otherwise each individual growth pattern is a possible norm or standard with all the other patterns possible deviations. Nothing consistent could result from such circumstances. Compiling an average increase produces a sample statistic plus error. If the average is all we know, then once computed, nothing more is forthcoming about growth. But the question remains as to how growth can be modeled to determine whether individuals follow or do not follow some consistent pattern. That is why Rasch specifies, for all organisms [of a consistent type] t_y must be the same. He was seeking a model by which to generalize growth, and only from such a condition can deviation be assessed and understood. In equation 2.3, Rasch specifies that y, given metameter t, is equal to the constant and rate from which data are derived.

Rao's example (1958) with children's weights follows a similar procedure to what Rasch described, and Rao mentions "metameter" several times in his paper. Metameter nowadays designates many attributes, but in this specific context, its definition can be stated as "a transformed value of a dose or a response, e.g. logarithm or probit, obtained by using a transformation equation that is independent of all parameters" (http://www.mijnwoordenboek.nl/definition).

The Rao equation is also simple and straightforward; one variable and two parameters describe the complexity of growth data. This simple expression should not hinder reflection on the fact that a structural equation remains the key. Furthermore, any model should be simple (i.e., parsimonious in its conception and predictive application), even for complex matters. A clearly specified conception is essential for validation studies. Even a simple regression equation can sometimes be difficult to fully implement, explain, and control.

Error e_i must also be considered where
$$Y_i = Y'_i + e_i = a + bX_i + e_i$$
and
$$\text{Variance } (Y) = \text{Variance } (Y') + \text{Variance } (e)$$

While random error can be modeled statistically, it is only defined and not completely prescribed/controlled, or else it would not be assumed random. There are numerous sources of error. Rasch (1977) frequently comments upon error as a concern in determining a growth model. One solution is to add more variables in an attempt to capture more information. Growth is complex so modeling its complexity with additional variables appears to be an option. But this opens Pandora's box to controlling/explaining further sources of error produced by employing more variables as well as producing additional problems such as multicolinearity. Introducing complexity sometimes only frustrates understanding.

The major issue, however, is not merely defining a regression equation whose constants are data dependent but determining a generalized growth model that is applicable, in this instance, to the development of piglets. This is a subtle but very important difference. Unless the regression equation is actually generalizable to all piglets, the goal of an encompassing growth law has not been achieved. Only when every individual regression equation, irrespective of sampling, is statistically the same can we approach success. This point cannot be overemphasized, and it is the reason that Rasch expressed dissatisfaction when consulting on studies that only produced differences without specifying a growth model.

Factor analysis or structural equation modeling (e.g., LISREL) uses a solution approach for dealing with an expanded structural equation, and some studies employ a theory for what should occur besides specifying initial conditions. In such applications, a simple regression expression expands to matrices to model this complexity:

$$Y = a + bX_i + e_i$$
$$\eta = B\eta + \Gamma\xi + \zeta$$

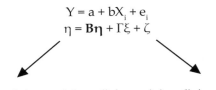

(m x 1) (m x m) (m x 1) (m x n) (n x 1) (m x 1)

But the form of the structural equation (for linking variables) and the magnitude of these variables still remain key issues to specify and determine experimentally. Proposing complexity is no improvement unless one can produce a viable growth model. A multivariate approach is no guarantee sui generis of success. Rasch (1977) alludes to this problem:

> The practical solution of this problem usually turns out to be rather easy by means of a theorem that has been presented (in Danish) in Rasch (Objektivitet, 1972), but as this is mainly a technical question we shall not enter into it here. (86)

The "main theorems" [i.e., theorems I and II discussed later] are given on pages 79ff. in Rasch's 1977 paper and are not shown here.

Another approach is suggested by T. Brody (1993). His strategy involves averaging over incidentals for bringing the result to a new, generalized level. Averaging over the incidentals moves data to a new and higher level of aggregation. It is similar to the problem of measuring one molecule versus measuring an aggregation of them. A single molecule's temperature is undetermined, but upon aggregation or accumulation, the sum swamps the incidentals. Simple averaging produces a new understanding of what could not be grasped singularly, but it is not without its problems. Another example is the comparison of the characteristics of a single word to that of the information summarized from the use of the same word in a variety of contexts. A single word seems less complex save for its initial presentation. With the advent of additional contexts, there are many more comparisons and applications possible. Comparing and contrasting these canonical applications brings utility and generalization to bear on the matter. However, it does not fully address the alternatives.

According to the ensemble approach as described by T. Brody (1993):

> A higher-level model built using these averages takes what
> above we called the irrelevant factors into account – jointly
> but not individually ... The problem of how to remove
> the unwanted information that depends on the initial
> conditions while keeping what does not so depend was
> solved by Einstein, and by Gibbs, using the method
> indicated above ... Averaging over this set of models
> technically known as an ensemble – is a trivial operation
> mathematically, but one that has the power of creating
> new concepts of quite different characteristics: ensemble
> averages no longer depend on initial conditions ...
> Probability is then nothing but a particular kind of average
> over the ensemble. (126)

The piglet example illustrates the $\log_{10}$ transformations. The growth rate averages are given by column (indicating the treatment code) at the bottom of the table together with the overall increase in weight for each piglet. Table 9.2 is Rasch's "Table 3" from his 1977 paper, which also includes a figure showing the logarithmic growth of each of the six piglets:

Table 9.2

[Table 3 as numbered in Rasch's paper; $\log_{10}$ transformations follow Rasch also.]

Logarithmic transform of the observed weights ($v_{it} = \log V_{it}$)
Age Logarithms base 10 of live weights: $v_{it} = \log V_{it}$
Growth mode in days

Days	KI,24	KI, 60	KII,6	KII,11	KIII,24	LIII,21	Log Average $v_{i.}$
20	0.60	0.68	0.65	0.71	0.70	0.68	0.672
25	0.69	0.80	0.79	0.82	0.83	0.82	0.792
30	0.79	0.87	0.90	0.91	0.93	0.91	0.886
34	0.88	0.94	0.96	0.98	0.98	0.96	0.952
38	0.95	1.00	1.02	1.02	1.04	1.03	1.010
42	1.01	1.08	1.10	1.08	1.11	1.08	1.078

48	1.10	1.18	1.17	1.17	1.18	1.18	1.166
53	1.18	1.23	1.25	1.25	1.27	1.24	1.235
59	1.23	1.28	1.29	1.28	1.30	1.26	1.274
64	1.22	1.32	1.34	1.35	1.36	1.34	1.336

Average 1.038

Table 2 Summary

Average

Growth 0.951 0.982 1.058 0.982 1.012 1.012 **0.965** rate β_t

Increase 0.62 0.64 0.69 0.64 0.66 0.66 **0.652** v_{it}

Some of the log values in Rasch's original table are slightly different from those computed and produced for this table, but the differences are slight, and they do not disturb the findings or comparisons. The differences are, perhaps, a result of previous hand calculations. Data points derived from the table are close to the totality derived from all six regression lines in Rasch's figure. Individual piglet regression lines are given in his paper. Some data points show deviation at each end of the regression line. Rasch (1977) draws his conclusions:

> As shown in Fig. 12.2 the said linearity holds - apart from small deviations that look unsystematic - in the whole experimental period for each of the 6 pigs. For an orientation we may then consider growth as a perfectly determinate process, thus neglecting the said variations, taking the straight lines connecting the first and the last point for each piglet as representing its exact growth process. (89)

Figure 12.1: Plot of average growth for six piglets from Rasch data

> Logarithmic growth of piglets during about 1½ months related to the average logarithmic growth of the whole batch of piglets. (Rasch 1977, 88)

Our figure 12.1 (Rasch's figure 2) shows the average growth for the piglets. As an aside, Gnanadeskian (1977), in explicating an example for data analysis, comments, "An interesting and seemingly appropriate transformation in this case would be to use logarithms of the original observations as the starting approach ..." (15).

The latent growth process generated from each piglet's data contributes to the growth trajectory function. If the piglets are assumed independent with respect to their genetic background, metabolism, eating habits, and so on, as described earlier, we also have a Brody trajectory in Rasch's analytical approach to growth.

Rasch presented each piglet's growth line to show individual growth patterns. These separate growth trajectories are not given here. To produce individual growth lines, each piglet column of measures is plotted against the average weight. Instead, we show their combination (ensemble), which clearly demonstrates Rasch's point of similarity in growth rate taking into account the terminal deviations previously mentioned. The fact that all the piglets in the experiment follow the same growth pattern is extremely important. Although each piglet has an expected unique trajectory, their statistical similarity implies a generic growth process independent of variations in feeding, or any other environmental considerations operating in the experiment.

By way of contrast, consider the economic growth example given in appendix A by Olsen (2002, 81–93). She acknowledges how varied the plots are for different countries frustrating a generality that might be applied to this data remarking (82), "the Growth Model does not apply to all countries at the same time." Some economists deem this problem the consequence of "anchovy goods" meaning that supply functions may vary extraordinarily due to seasonal or other environmental conditions. We expect individual differences, sometimes seasonal differences, but only within the context of a general model if seeking to establish a growth pattern. We might expect the same variation in human growth. Pediatricians may consult a standardized infant growth chart for weight and length for their patients, but each infant or child must be considered uniquely.

Rasch elucidates the comparing function, but he also indicates its limits:

... parameters of (VIII:3) are identified with empirical values

$$T_t = \upsilon_{.t} - \upsilon_{.0}, \quad \beta_i = \frac{\upsilon_{i44} - \upsilon_{i0}}{\upsilon_{.44} - \upsilon_{i0}} \quad \alpha_i = \upsilon_{i0} \qquad \text{Rasch, VIII:4}$$

However unambiguous this result is, it does not permit specifically objective comparisons between the parameters for two pigs by means of their weights at the same time; as a matter of fact there exists no comparing function of

$$v_{it} = v_{i0} + \beta_i \tau_i \qquad and \qquad v_{jt} = v_{j0} + \beta_j \tau_j \qquad\qquad Rasch, VIII:5$$

which is independent of the age parameter τ_i. - Nor do the weights of the same pig at two ages allow a comparison of the two age parameters which is independent of the parameters of the pig (89).

Rasch indicated there is no "comparing function" in this data frame, and hence, no "specific objectivity" for the parameters of any two piglets by (1) weight or by (2) age because neither parameter is independent of the parameters of the piglets. He continues:

> ... proper subject of the study is not the single weights, it is their growth, ... *the progress of weights ...*
>
> from age t_1 to age t_2 has something to do with the size of the pig at the first age: *accretion of weight at a given time t occurs by way of production and destruction of cells, assumption of proportionality for this relationship will lead to an attempt at expressing growth as a relative increase, that is by means of the ratio increase in v_t as defined by (VII:2) can be expressed as a transformation of the **relative increase** ... by means of the ratio*

$$\frac{V_{t+1} - V_t}{V_t} \qquad Rasch, VIII:6$$

In Rasch's data, the mean piglet slope is b = 0.014 with slight variation among the six piglets' individual slopes, but differing only in the fourth decimal. R^2 narrowly ranges .97–.98 in a linear model for all piglets. Given Rasch's stipulations, we might conclude there is a consistent growth pattern among the piglets from these ensemble statistics. The data plot shown above indicates little variation among the treatment conditions. While a variety of independent variables and conditions might be hypothesized as contributing to piglet growth, there is no indication from the plot that any remarkable deviations are evident "independent of the age parameter." A one-way ANOVA was computed across columns with F = 0.293 (p = 0.915 for 5, 54 df). Statistical significance requires a critical value of 2.386 at .05, so these column differences are clearly not statistically significant. The average growth pattern appears similar for all six piglets. The ensemble values are generally stable with very similar data plots.

Growth values are given in table 3: Increases in piglet growth. They indicate the average gain in weight by day with ln and $\log_{10}$ transformations of weight, weight difference, and cumulative increase in weight.

Table 9.3

Increases in piglet weight

Day	Average Weight	ln Average	log Average	Weight Difference
20	4.70	1.548	0.672	
25	6.20	1.825	0.792	1.5
30	7.70	2.041	0.886	1.5
34	8.96	2.193	0.952	1.26
38	10.23	2.325	1.010	1.27
42	11.96	2.482	1.078	1.73
48	14.65	2.684	1.166	2.69
53	17.18	2.844	1.235	2.53
59	18.80	2.934	1.274	1.62
64	21.66	3.075	1.336	2.86

The average gain per measurement was 1.884. There appears to be a change in weight difference after day forty-two. The average gain was 1.382 for the first four measurement periods, and 2.286 for the last five measurement periods. Column differences for the first four measurements were not statistically significant ($F = 0.427 < 2.57$), and the same was true for the last five measurements ($F = 0.389 < 2.37$). The minor differences observed in the deviation of data points about the regression line in figure 1 are not in conflict with the ensemble value.

We cannot tell for sure what is producing growth by a simple weight increase. It could be systematic or random. To determine whether or not these changes occurred by environmental effects, the piglets' genes, or other factors would require further experimentation. We also recognize growth is not continuous and linear. Plateaus interspersed by (discrete) growth spurts are notable in children and adolescents. Sometimes large increases are observed. No coefficient is ever assumed applicable over the totality of growth. Variation in differences is frequently evident in the early and late stages of human life. Sometimes they are the consequences of genetic and environmental advantages or disadvantages. We cannot tell from the data alone what is producing this change. It could be a random effect. Perhaps it is something unrecognized. To determine whether or not this change was due to an environmental effect, the piglets themselves or other factors would require further experiments to evaluate these possibilities.

We give Rasch's (1977) generalizations for the piglet study:

1. Specific Objectivity.

The question of *possible specific objectivity of such comparisons answered through modifications of the theory of V-VII and modifications of the main theorems.*

> ... generalizations to complex processes, with interactions of several factors, with states that are influenced not only by the immediate preceding state, but by several or all previous states.

$$\Upsilon_{t+1} - \upsilon_t = \log \frac{V_{t+1}}{V_t} = \log \left(1 + \frac{V_{t+1} - V_t}{V_t}\right) \qquad \text{Rasch, VIII:7}$$

Main Theorem I states that specifically objective comparability of pigs as well as of age levels in a scalar parameterization with respect to growth requires the reaction parameter, i.e., $v_t + 1 - v_t$, to be latently additive in terms of one parameter for the pig and one for the age level. That this is actually the case follows from (VIII:3) since

$$u_{i't+1} - u_{i,t} = (u_{i0} + \beta_i \tau_{i+1}) - (u_{i0} + \beta_i \tau_i) = \beta_i \delta_i \qquad \text{Rasch VIII:8}$$

where $\delta_t = g_{t+1} - g_t$ ... and (VIII:8) is multiplicative, thus latently additive (90).

The reaction parameter, weight gain, is a function of the feeding protocol. A transformation of the increase is described next. Rasch refers to Theorem I to substantiate his next steps.

2. Growth Model.

The growth process defined by (VIII:8) is a special case of a class of determinate dynamic systems where every object O_i of a collection goes through a series of states in which they are confronted with a sequence of agents $A_1, ..., A_{i'}, ...$ assumed each object is characterized by a permanent parameter b_n and initially in a state O_{n0} with a parameter value x_{n0}. Under influence of the agent A_1 with the parameter value α_1 the state is changed from O_{i0} to O_{i1} with $\omega_{i'}$ in a relationship which may be written

$$\xi_{i1} = \varrho \, (\xi_{i0} \mid \omega_i, \alpha_1) \qquad\qquad \text{Rasch, VIII:9 .}$$

This function ... the *transition function* α_2, changes ξ_{i1} into

$$\xi_{i2} = \varrho \, (\xi_{i0} \mid \omega_i, \alpha_1)$$

where ξ_{i1} is given by Rasch, VIII:9. $\hspace{3cm}$ (90)

3. Referential System.

Two sorts of comparisons exist:
(1) Between permanent parameters of different objects on the basis of their state parameters at the same two or more stages of the process.
(2) Between two agent parameters on the basis of two or more state parameters for the same object.
(3) Comparisons between the initial states of two objects on the basis of later states of the objects do not give rise to specific objectivity (90).

The statements by Rasch concerning two types of comparisons have been numbered for clarity. Comparisons can be made by piglet from two or more stages by measurement of their weights (1). Likewise, comparisons can be made for treatments by column comparisons (2). Rasch indicates that initial states of objects do not attain specific objectivity by virtue of later states (3). This is important for understanding exactly what is implied from making comparisons; for what does and does not lead to specific objectivity, and to what is meant by his specification of an "individual-oriented" psychometrics.

Rasch speaks to the frame of reference in *Section IX Indeterminate Frames of Reference.*

> (1) In IV-VIII the exposition has been concentrated on determinate frames of reference. (2) In analysis of empirical data this at best means an idealization of the situation, of both the general gas equation and the growth of pigs gave rather precise descriptions of the actual data. ... the physical example contains measurement errors and the biological case also capricious variations in the digestive processes. (91)

The structural equation must include an error term—which is Rasch's point. The common problem frustrating derivation of valid information from data is error. Error lies with the data. Rasch (1977) strives to explicate theory and probability to counter this problem.

> In physics ... "genuine" uncontrollable responses, e.g., radioactive radiation in atomic theory governed by t random events - not totally lawless, ruled by probabilities inherent in the models. ... The mapping of all indeterminacy of responses on the concept of probability - which actually is derived from "mere toys" like games of roulette, dice and cards - is today not only prevalent in large areas of physics, but also in biology and in social science. (91)

The previous statement by Rasch indicates that a scientific outcome is to be governed by theory and not by data. Generality and "laws" arise from the establishment of outcomes governed by theory modeled by data. It is not the data we focus upon because data is error prone and fallible. It is the substantiation of theory illustrated by data that is demonstrated or not via experimentation.

Rasch's comments on physics, indeterminacy, and probability return us to Brody, who was seeking to take data to a higher level by virtue of abstracting the governing law without the confusing aspects of data error from experimental outcomes. In a structural equation model, the initial values and form are not known. They are hypothesized from another study (borrowing error) or surmised without a basis for their use. Only a well-conceptualized construct theory (ignoring error) could supply the values to be tested.

Any productive solution requires interplay between the theory/ hypothesis and the nature of the structural equation model. Good theory is necessary because statistics alone cannot produce valid results. A dynamic interplay between theory and equation modeling is the hallmark of scientific work. Rasch employed equation VIII: 3 with data collected from a field experiment in animal husbandry to derive a *growth model*. Rasch (1977) speaks in the quotes given earlier and those forthcoming of the need to bring valid data to the level of law for "objectivity" to be forthcoming.

For the concept of specific objectivity to be of use, it is therefore essential that it can be applied to indeterminate—more concretely: to probabilistic—frames of reference.

> ... A treatment of this topic corresponding to the treatment in IV-VIII of determinate frames of reference must await a later occasion. (91)

> ... it was possible to map this indeterminacy on the Poisson probability distribution (I:1), and its parameter was in I,4 shown to be decomposable into a product of a factor for the student and a factor for the test, expressed in the formula (II:1). (91)

Feller (1940) offered such an exposition, *On the logistic law of growth and its empirical verification in biology*. His purpose was to apply the logistic model to biological data on growth. The logistic curve provides a graphical representation of the logistic equation proposed in 1838 by Verhulst whereby

$$\frac{dN}{Dt} = aN\,(M - N)$$

describes the growth of N, and a represents growth dN/dt proportional to N (size in time t) and what remains to be expressed by growth. The integral of the differential equation is

$$N = \frac{M}{1 - e^{-a(t-t_0)}}$$

Coefficient a is the growth capacity at t_0 the time when growth is half of its maximum. *Characteristic time* designates the interval from 10 percent to 90 percent of maximum size M.

The logistic curve

$$Y(t) = \frac{\alpha}{1 + e^{-\beta(t-m)}} \qquad (x > \alpha, \beta > 0)$$

and corresponding differential equation

$$\frac{dy}{dt} = \beta y \left(t - \frac{y}{\alpha}\right)$$

with α, β, m constant for the characteristic/specie is one of several variants to the model.

A small example given in table 9.4 using shrimp data serves to illustrate the process.

Table 9.4 Shrimp growth data

Age	Growth	Logistic Model
0	0.24	0.37
1	2.78	2.51
2	13.53	13.62
3	36.30	36.30
4	47.50	47.42
5	49.40	49.55

A logistic model of these observations with shrimp data growth is shown in figure 9.2. The plot could also be presented as a straight line by taking logarithms.

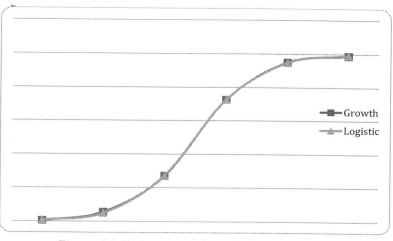

Figure 9.2 Shrimp growth and a logistic curve

The logistic curve (Feller 1940) proves an excellent fit to the growth data in this example. The curves overlap, and the correlation between the observations and values taken from the logistic is 0.99. Transforming the observed growth data by $\log_{10}$ provides a correlation of 0.92 to the observations. The $\log_{10}$ transformation of the observed data and the logistic equation correlate 0.88. This correlation is somewhat lower, but when the upper and lower data points are eliminated, the correlation increases to 0.96. This higher value was produced because the lower and upper tails of the sigmoid curve were eliminated, bringing the intermediate values to greater prominence with a constant increase in the correlation coefficient. The tails of a distribution are almost always extreme values, frequently the most deviant, and are not brought into exact correspondence by log transformations. This deviation of tails was observed earlier with Rasch's $\log_{10}$ transformations of the piglet growth observations on weight, which indicated some differences at the ends of the regression line. Logistic transformations do not linearize the tails exactly (Rasch 1960). *Characteristic time* Δt has been designated as the time needed to grow from 10 percent to 90 percent of maximum size, providing a more stable estimate of the growth process, and this reduction of range effectively eliminates the deviations frequently encountered in the tails. An excellent fit of the model to the data does not validate the data or the model. It simply permits the model to speak for the data without the encumbrances of problems in data procurement and sampling.

The Knox Cube Test-Revised (KCTR, Stone 2002b), a short test of visual attention and short-term memory, offers another example, shown in figure 9.3.

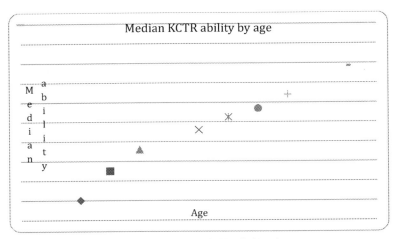

Figure 9.3 Median KCTR ability by age

A logarithmic model fits the data very well for a sample of slightly more than 1,900 persons. This was a subset of the total sample ranging from age two to ninety for more than three thousand persons. This data is cross-sectional and not longitudinal. While the KCT is an individually administered instrument useful for measuring individual growth, it was administered to a large sample where the maximum ability of the average person was reached by age twelve, although others did even better and performed higher over this age span. The logarithmic growth curve is steepest between ages two and twelve. After age twelve, the curve tapers off to a very modest increase through adulthood until age fifty for the median individual. From that point on, the curve turns slightly downward, with a greater deviation found among elderly individuals. They may be exhibiting cognitive impairment from a variety of conditions. It is clear that growth varies with age (time).

The trajectories of individual growth may vary due to a variety of conditions related to genetic integrity or pathology, environmental conditions, and the stage of growth. While these conditions may affect individuals, we may still seek to investigate general growth conditions. Accurately determining individual and general growth trajectories is a very complex matter.

A Reading Example

The transition from growth models for animal husbandry to applications in reading development is straightforward. We illustrate this in figure 9.4 showing a web (eliminating the connecting lines) of twenty Lexile values produced by twenty randomly selected students participating in a longitudinal reading study involving several thousand students.

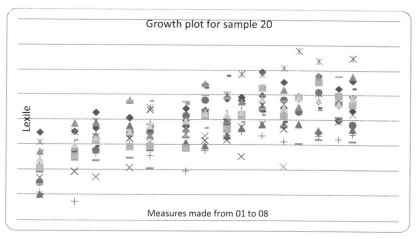

Figure 9.4

Figure 9.4 is a composite plot of the individual reading measures for each of the twenty students. A personal trajectory is important in understanding the progress of each student. A generalized model of growth following Rasch's analysis for piglets helps unravel the network of individual trajectories into a single trajectory of mean growth increase expressed in Lexiles for time 2001 to 2007. The growth model assists in understanding the progression in reading ability over time. This is shown in figure 9.5.

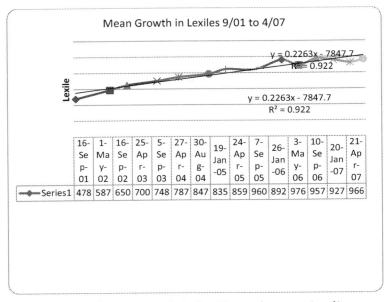

Figure 9.5 Mean growth in Lexiles and regression line

Figure 9.5 gives a plot of mean values, and table 9.5 provides the pre- to postassessment statistics. Rao (1958) argued that mean values taken at points in time can be used to indicate growth. It is a less efficient procedure for generalizing growth than employing a more powerful model, but Rasch (1977) indicated that individual values did not describe a growth model.

Table 9.5

Sample 20 Pre- to Post-assessment Statistics

	16-Sep-01		21-Apr-07
Mean	478.2	Mean	966.8
Median	408	Median	953
Standard Deviation	209.3	Standard Deviation	192.8
Range	704	Range	703

There was an average increase of 488.6 Lexiles from a preassessment of 478.2 to the postassessment of 966.8 Lexiles over the six-year period. This increase between preassessment to postassessment is as great as the initial preassessment differs from a Lexile of virtually zero to the level of the preassessment. The regression slope is b = 0.226 for the total sample while individual slopes ranged between 0.097 – 0.398 around this center shown in figure 6.

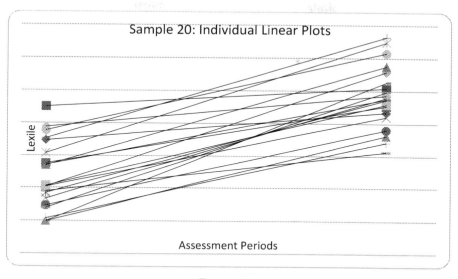

Figure 9.6

In Rasch's piglet example, the dependent variable was log-weight regressed on days. In the reading example, it is the Lexile measure of reading ability regressed on time. Log-weight served as the continuous variable for the piglets over the time span, while the Lexile reading measure provided a uniquely continuous variable required for measuring reading ability over a wide range of reading ability measures spanning six years.

Using a common measure over the entire range of possible values obviates difficulties that occur when different instruments are required to address the span of respective values often necessitating a linking of instruments. We might take piglet weight for granted as a continuous dependent measure over any span of time, but even well-fattened hogs reach a maximum that establishes a practical limit for this outcome measure.

Measuring reading ability in multiyear educational applications frequently rests upon a design of administering one test form for the initial assessment and different form for each successive year. The Lexile Framework (Lexile, 2000; Stenner, Horabin, & Smith, 1989 is applicable over the entire six-year time span and beyond providing a continuous measure (exceeding even the measures required for this sample spread) without necessitating any attempt to statistically connect two or more different test instruments as is frequently required for a longitudinal study (Williamson 18; 16; 15).

While consistency in a measuring instrument is essential, so is student consistency in attendance. A highly mobile population has been typical in the United States. This mobility has an influence upon consistency in schooling. A large number of students change schools as their parents relocate, separate, and divorce. Continuous entry and leaving a school system is not conducive to educational progress. A high student dropout rate, typical in urban locations, also means that assessments are incomplete.

Growth in reading for this sample occurred as a consequence of individual influences and instructional factors over a period of about six years. The average reader progressed from text title examples such as *Harold and the Purple Crayon* (490 Lexiles) to *Leon's Story* (970 Lexiles). The standard deviations for these two assessments were similar; SD = 209.3 to 200.6. The minimum-maximum spread was 197-901= 704 Lexiles for the initial assessment and 625-1328 = 703 Lexiles for the final assessment. An R^2 of .96 for the student's Lexiles indicates this linear equation explains most of the variance in the sample data.

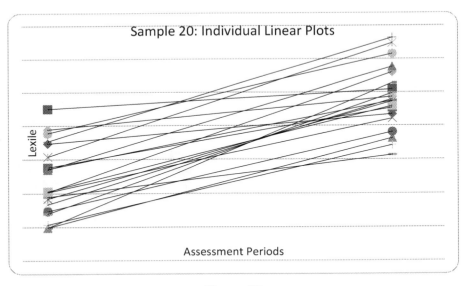

Figure 9.7

Figure 9.7 shows the individual linear plots for each of the twenty students in the random sample. Some readers are below grade level. Others are approaching college level. The range of values at the pre- and postassessments was greater than the average growth over six years. These large individual differences indicate the wide range of reading competency that exists among classmates in public schools and typical of what teachers must continuously face by the varied reading competence of students. Figure 9.8 shows the linear plots for the seven students achieving the highest measures in the last assessment. Six of the seven show a somewhat similar rate of growth. One student was the highest of the seven in the first assessment but scored lowest of these seven in the final assessment. Such an anomaly is not uncommon. Reading growth can be influenced by a variety of factors. One major contributor is the interest level and attention demonstrated in daily work and on assessment tests. Attention may vary by subject matter interest. Anomalies in reading progress sometimes identify the noninterested student, the lackadaisical one, and sometimes signals that home-life conditions are not as they should be.

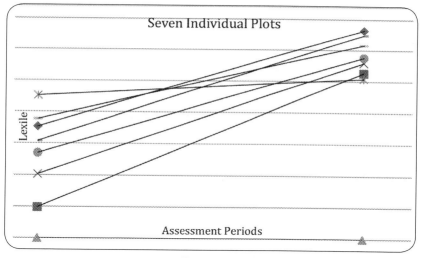

Figure 9.8

Reading ability was well captured by the Lexile Framework, but there is no easy solution to accounting for the overall growth, or the personal growth factors governing every student's attitude to reading, or learning in general. Students are powerfully influenced by personal and family values even assuming equality of opportunity and schooling. The totality of these factors may produce large individual differences. But even with such assessment variability for individual students, there are clear indications of overall growth in reading ability for this sample that we might infer for the population they represent. Figure 9.8 shows the individual plots for seven students exhibiting the highest growth rates. Note the highest performing student in the first assessment is the lowest in the final assessment among these seven.

Figure 9.9 gives individual plots for the remaining thirteen students.

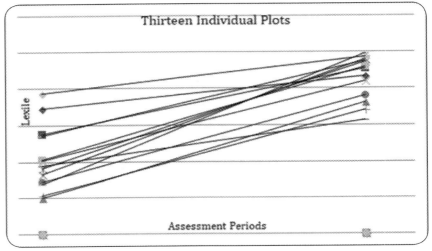

Figure 9.9 Thirteen individual plots

These plots all show varieties in rate of growth not uncommon to observe in public schools, although there is a positive growth rate for all of them.

The richness from this approach to determining growth rate is that not only can we produce growth trajectories for individuals, groups, and subgroups, we can introduce predictions by extrapolating these growth rates forward and backward. We can set expectations for future progress against actual growth and evaluate its occurrence. We can also look backward at what might have transpired in the past leading to the current status of the student, such as consistent schooling versus varied attendance, new student versus dropout, and so on.

Exploratory analyses using regression are basic, simple, and useful (Rogosa, Brandt, and Zimowski 1982; Rogosa and Willet 1985). They permit change to be investigated and growth parameters to be estimated. Rogosa, Brandt, and Zimowski (1988) offered ten *Mottos for the Measurement of Change*. The first three speak to this occasion:

1. Individual time paths are the proper focus for the analysis of change.
2. A model for individual change is useful for the measurement of change.
3. The collection of individual (X on t) regression functions is the key initial summary of the data. (744)

Summary

In Rasch's piglet example, log-weight provided a continuous variable regressed on time, while the Lexile measure of reading ability provided a continuous variable on time for modeling reading growth. Weight and Lexile measure provided an important service giving substantive meaning to the respective values for denoting the outcome of growth. Using a constant measure over the entire range of possible values obviates difficulties that occur if different instruments are required to address the span of respective values. Piglet weight stands for the totality and pattern of piglet growth. Because Lexiles eliminate any statistical adjustment to connect different test instruments it is ideal for the kind of longitudinal study described here.

We find it especially interesting to investigate Rasch's involvement in the matter of growth in contrast to his popular association with psychometrics although we find a connection not discussed here. Olsen (2002) provides some interesting remarks on these matters, including the trip Rasch made to India in 1951. C. R. Rao was among the participants. His paper (1958) on time series models had direct impetus from this visit, and Rasch is acknowledged in this paper. According to Olsen (2002, chapter 2) a large amount of miscellaneous notes remain unpublished on these matters. Olsen (2002) also addresses Rasch's work on growth models applied to econometric matters, accidents, and so on that are also not discussed here. Rasch appears an early advocate of log transformation to enhance data analysis. Application of stochastic processes to these other areas are given by Olsen, who includes important information provided by Allerup, Andrich, Jensen, and Stene in chapter 2, "Georg Rasch's Growth Models." While Rasch's work on growth models has been eclipsed by the application of his ideas to psychometrics, we have examined Rasch's example with piglets to derive the essence of his ideas, and utilized an example on reading ability to further illustrate Rasch's life-interest in growth models.

Box 30: Individual-Centered Measures
versus Group-Centered Measures

The preface to *Probabilistic Models for Some Intelligence and Attainment Tests* (Rasch 1960) cites Skinner (1956) and Zubin (1955) in arguing that "... individual-centered statistical techniques require models in which each individual is characterized separately and from which, given adequate data, the individual parameters can be estimated" (Rasch 1960). Both authors are cited for their advocacy of individual measures, and while the Skinner reference is easily located, the mimeographed work by Zubin has not been found. Rasch goes on to state, "... present day statistical methods are entirely group-centered so that there is a real need for developing individual-centered statistics." What constitutes the differences in these statistics?

While individual persons and groups of persons are the focus of this discussion, we begin with a simple illustration because human behavior is complex and easier to illustrate with a single variable. We choose temperature for this illustration because the measuring mechanism for temperature is well established by a single measure of degree, disregarding windchill, altitude, and so on (Stenner, Stone, and Burdick 2004). A measuring mechanism consists of (1) a substantive theory, (2) successful instrument fabrication, and (3) demonstrable data by which the instrument has established utility in the course of its developmental history formulated by theory.

Consider six mercury-tube outdoor thermometers that are placed appropriately in a local environment but near each other. They all register approximately the same degree of temperature, verified by consulting NOAA for the temperature at this location. One by one, each thermometer is placed in a compartment able to increase/decrease the prevailing temperature by at least ten degrees. Upon verifying the artificially induced temperature for each thermometer, it is returned to its original location and checked to see if it returns to its previous value and agrees with the other five.

If each of the six thermometers measured a similar and consistent degree of temperature before and after the artificially induced environmental condition, this consistency of instrument recording demonstrates a case for reliability. Each thermometer initially recorded the same temperature, and following a change to and from the artificial environment returned to the base degree of temperature. Furthermore, all the measurements agree.

Interestingly, the experimentally induced change of environment produced what may be called *causal validity*. The temperature was manipulated between the artificial and natural environments. When measuring mechanisms such as outdoor thermometers are properly manufactured, this result is to be expected, and this experimental outcome would be predicted prior to environmental manipulation from all we know about temperature and thermometers. This outcome might further be termed *validity as theoretical equivalence* (Lumsden and Ross 1973) because the replications produced by all six thermometer recordings might be considered "one" temperature, give or take errors of measurement. Our theoretical prediction is expected as the consequence of the causal process produced by the experiment, and reported by all the instruments. *Causal validity* is a consequence of the measuring changes that can be produced in an experiment. Its essence is "prediction under intervention." The manipulable characteristics of our experiment involving the base environment, change made by way of an artificial environment, and the final recorded temperature are the consequences of a well-functioning measuring mechanism. Each of the six individual thermometers records a similarly induced experimental deviation and a return to the base state. Each thermometer constitutes an individual unit, and the six thermometers constitute a group, a unit, which is exactly what would be predicted.

Now consider a transition to a physical human characteristic. Height is the new outcome measure, and the determination of height at a point in time can be obtained from another well-established measuring mechanism—the ruler, which provides a point estimate for one individual measured at a point in time. When this process is continued for the same individual over successive time periods, we produce a trajectory of height for the person. From these values, one may determine growth over time intervals as well as any observed plateau (well known to occur in individual development). The individual's trajectory rate may also vary because of illness and old age, so we could discover different rates over certain time periods as well as determine a curvilinear average to describe the person's total individual trajectory. Growth in height is a function of time (duration) and the human characteristics resulting from genetic and environmental makeup. These statistics are intra-individually determined. Such a statistical analysis establishes the "individual-centered statistics" that Rasch spoke about earlier.

Combining individual measurements of height into a group or groups is a common method for producing "group-centered statistics," often employing some frequency model such as the normal curve. This is most common when generalizing the characteristics of human growth in overall height based upon a large number of individuals. The difference between measuring a group of individuals compared to our first illustration using a group of thermometers is that while we expected no deviation among the thermometers, we do not expect all individuals to gain the same height over time but register individual differences. Hence, we resort to descriptive statistics to understand the central trend, and the amount of variation found in the group or groups. An obvious group-centered statistical analysis might be made by gender, comparing the height of females to males.

The measurement of height is straightforward, and the measurement mechanism has been established over centuries. The same cannot be said for the measurement of mental attributes occurring in psychological and educational investigations. Determining the relevant characteristics for their measurement is more difficult, although the procedures for their determination should follow those already discussed. The major statistical hurdle is moving from the ordering of a variable's units to its "measurement application." The measurement model of Georg Rasch has been instrumental in driving this process forward.

Do we know enough about the measurement of reading that we can manipulate the comprehension rate experienced by a reader in a way that mimics the above temperature example? In the Lexile Framework for Reading (LFR), the difference between text complexity of an article and the reading ability of a person is causal on the success rate (i.e., count correct). It is true that short term manipulation of a person's reading ability is, at present, not possible, but manipulation of text complexity is possible because we can select a new article that possesses the desired text complexity such that any difference value can be realized. Concretely, when a 700L reader encounters a 700L article, the forecasted comprehension rate is 75 percent. Selecting an article at 900L results in a decrease in forecasted comprehension rate to 50 percent. Selecting an article at 500L results in a forecasted comprehension rate of 90 percent. Thus we can increase/decrease comprehension rate by judicious choice of texts—that is, we can experimentally induce a change in comprehension rate for any reader and then return the reader to the "base" rate of 75 percent. Furthermore, successful theoretical predictions following such interventions are invariant over a wide range of environmental conditions, including the demographics of the reader (male, adolescent, etc.) and the characteristics of text (length, topic/genre, etc.). This outcome is the consequence of years of research involving a measurement mechanism that produces consistent results.

Many applications of Rasch models to human science data are thin on substantive theory. Rarely proposed is an a priori specification of the item calibrations (i.e., constrained models). Causal Rasch models

(CRM) prescribe (via engineering, manufacturing, fabricating) that item calibrations take the values imposed by a substantive theory (Burdick, Stone, and Stenner 2006; Stenner, Stone, and Burdick 2009; Stenner and Stone 2010). For data to be useful in making measures, it must conform to the invariance of both Rasch model requirements and substantive theory requirements as represented in the theoretical item calibrations, two attributes that we specify as *doubly prescriptive*. When data meet both sets of requirements, the data are useful not just for making measures of some construct but also for making measures of the precise construct specified by the equation that produced the theoretical item calibrations. Doing so distinguishes a causal model from a descriptive one.

A causal (constrained) Rasch Model that fuses a substantive theory to a set of axioms for conjoint additive measurement (Stenner, Burdick, & Stone, 2008; Stenner, Stone, & Burdick, 2009) affords a much richer context for the identification and interpretation of anomalies than does an unconstrained Rasch model. First, with the measurement model and the substantive theory fixed, it is self-evident that anomalies are to be understood as problems with the data, ideally leading to improved observation models that reduce unintended dependencies in the data. Second, with both model and construct theory fixed, it is obvious that our task is to produce measurement outcomes that fit the (aforementioned) dual invariance requirements. An unconstrained model cannot distinguish whether it is the model, data, or both that are suspect.

Over centuries, instrument engineering has steadily improved to the point that for most purposes "uncertainty of measurement," usually reported as the standard deviation of a distribution of imagined or actual replications taken on a single person, can be effectively ignored. The practical outcome of such successful engineering is that the "problem" is virtually nonexistent (consider most bathroom scale applications). The use of pounds and ounces also becomes arbitrary as is evident from the fact that most of the world has gone metric. What is decisive is that a unit is agreed to by consensus and is slavishly maintained through use of substantive theory together with consistent implementation, instrument manufacture, and reporting.

We specify these stages:

Theory ⟶ Engineering ⟶ Manufacturing ⟶ Quality control

These are the steps by which measures are consistently produced and continuously validated. The doubly prescriptive device evaluates the quality of the process.

Different instruments qua experiences underlie every measuring mechanism, such as environmental temperature, human temperature, children's reported weight on a bathroom scale, and reading comprehension. From these illustrations and many more like them, we determine point estimates and make comparisons for individuals or for a group. This outcome lies in well-developed construct theory, instrument engineering, and manufacturing conventions that we designate *measuring mechanisms*.

Understanding the Basis for a Unit Requires Consideration of Counting and Measuring

Still further, consider these four words as paired: *counting versus measuring*, and *qualitative versus quantitative*. Closely allied to these pairs are two more related words: *continuous and discontinuous (discrete)*. Any consideration of these six words, their foundation, and implications would have to be extensive and will be limited here to what these words imply in a Rasch context for understanding a unit in counting and measuring.

These six words encompass a complex but important problem. Good (1983, 157) says, "... all science and all language is based on a swamp." There is truth to this comment seemingly due to frustrations in dealing with human behavior, but persistence in the pursuit of clear and distinct ideas may prove of value. How Rasch might have explained each of these words can only be implied from his writings because he did not define them explicitly, although he did use them in ways that can be examined. He employed two additional ones needed for assistance and clarification of the six already identified words: *difference* and *analogy*. Each of these eight words needs to be explicated and integrated to make clear what it means to count or number, and what it means to measure.

We begin by inquiring of the properties that align and (or) distinguish numbering and measuring. Richard Dedekind (1963) did so in the preface to *Was sind und sollen die Zahlen?* [What is the meaning and purpose/use of numbers?] He answered his own question:

> My answer to the problems propounded in the title of this paper is, then, briefly this: Numbers are free creations of the human mind; they serve as means of apprehending more easily and more sharply the difference of things. It is only through the purely logical process of building up the

science of numbers and by thus acquiring the continuous number-domain that we are prepared accurately to investigate our notions of space and time by bringing them into relation with this number-domain created in our mind. If we scrutinize closely what is done in counting an aggregate or number of things, we are led to consider the ability of the mind to relate things to things, to let a thing correspond to a thing, or to represent a thing by a thing, an ability without which no thinking is possible. Upon this unique and therefore indispensable foundation, as I have already affirmed in an announcement of this paper, in my judgment, the whole science of number be established. (31–32)

First, we need to cognitively discern *differences between things,* as Dedekind puts it. Gross differences are usually quite easy to determine. Dedekind indicated that numbers serve to discern more precise differences. He said that the science of numbers allows "things" in space and time to be scrutinized. That is because things and numbers are brought into a complementary relationship. Furthermore, a number can correspond to a thing analogously by providing the *indispensable foundation* that Dedekind prescribes in his remarks. He suggests a way to build a foundation in spite of the swamp of language.

Just as Dedekind indicates that numbers are creations, so too are observations. They are the "things" we fixate upon in contrast to the ones we ignore. We do not "find" observations or "collect" data but "produce" them. We count apples and ignore all else until we are done.

Numbers and things for Dedekind are paired in a way that may be illustrated as:

Numbers

Things

Relationship is 1:1
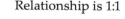

The relationship between things and numbers is one to one for Dedekind. But when each thing differs from the next thing, this difference produces

291

the unit for things (be they apples, houses, etc.). Each count also differs from the next one, beginning with zero, by the unit of number expressed as "one," a positive integer. Things differ as produced by systematic observation, but Dedekind in paragraphs #72 and #161 indicates they do so precisely when rational enumeration drives the difference. Numbers may signify collections of things because things can be differentiated and enumerated analogously. Dedekind indicates when numbers express a determinate quality, or when the things are counted and set in order and appear as ordinal numbers. When an aggregate of things is arranged along a linear dimension, things can also be enumerated, sequenced, and numbered. In this context, numbering and counting appear synonymous. It is the things and numbers we physically arrange or "think about" when contemplating things and numbers.

Arranging and numbering things can only be done with things and numbers that are clear and distinct so one can discern their differences. Five things added to two things gives seven things. Eight things minus three things gives five things. This is one of the advantages that Dedekind ascribes to the use of numbers for things. Numbers become the analog of things when we are in this mode of thinking. But we should never lose track of the goal in science, which is to discern the properties of things with more clarity. Numbers help to bring this about when investigating things.

Things are discontinuous units. They are separate entities such as apples, houses, and so on. Reason makes them different, not simply "pure" observation. Our eyes alone do not separate things. Our cognition does it. Numbers are entities also because they too are distinguishable, but numbers are sequenced and continuous as when we first learned to count (1, 2, 3, etc.). Numbers enumerate things, and a count of things is the sum derived from this enumeration. Numbers and things can be added or concatenated, and subtracted.

Someone may say this resembles Steven's (1936) nominative and ordinal levels of scaling, and this is true. We enumerate using some alphanumeric system such as names, passwords, combinations, and phone numbers, all of which illustrate this process. However, the crux of the matter is how we imply or infer properties about things and what they consist of. If we follow Dedekind's strategy, things can be numbered and dealt with accordingly. Things may be manipulated by arithmetic. We do not change the thing itself, but we can manipulate their numerical analogues and arrive at new insights about the properties of things. Bergmann and Spence (1944) write,

> Loudness, hotness, length can be differentiated ...
> Properties of physical objects or events which are amenable
> to such differentiation we shall call physical dimensions.
> Physical measurement consists in the assignment of
> numbers to the objects or events of a physical dimension in
> accordance with certain rules (laws and conventions). (6)

Runners may be distinguished by their numerals, but they may be further distinguished in their ranking or ordering by skill. Marathon runners are distinguished by their numerals, the order in which they cross the finish line, and by the time recorded for their feat. A specific marathon runner (a thing) is expressed as an analog by (1) a numeral designation, (2) an ordinal position when crossing the finish line, and (3) an elapsed time in hours, minutes, and seconds. The last level is clearly a continuous measure of duration as officially recorded. Continuous duration becomes an analog for a runner by using an analog stopwatch or digital recording device. The latter device, now commonplace, reminds us that numbers can be expressed as discontinuous by their succession but continuous when seeking only the duration or sum from a beginning usually set at zero. Rasch (1977) indicates,

> That science should require observations to be measurable
> quantities is a mistake, of course; even in physics
> observations may be qualitative – *as in the last analysis
> they always are.* (2, our emphasis)

Numbering would be completely satisfactory as long as enumeration sufficed for our needs. Numbering is "measuring" in this context when the count of things is all we require, but this usage implies that things be treated as equal and equivalent. In this context, one thing is fully exchangeable for another. However, to advance beyond enumeration, more understanding is required, as in for example the study of spelling. If words-spelled-correctly is a sufficient index, then enumeration is satisfactory, but this assumes all spelling words are equal and exchangeable. How shall we account for equating the words cat, louse, and rhinoceros? Thorndike (1904) indicated:

> If one attempts to measure even so simple a thing as
> spelling, one is hampered by the fact that there exist no
> units in which to measure. One may arbitrarily make up
> a list of words and observe ability by the number spelled
> correctly. But if one examines such a list one is stuck

by the inequality of the units. All results based on the equality of any one word with any other are necessarily inaccurate. (7)

We can count words (units) spelled correctly, but we cannot measure spelling ability. While there is a difference between the count of the words spelled correctly and the measuring of spelling ability, in casual usage the two are frequently equated. Measuring spelling ability is hampered by (1) lack of a reproducible unit, (2) reliance upon the sample of words, and (3) further reliance upon the sample of persons. These problems must be resolved to make clear the difference between counting words spelled correctly and measuring spelling ability.

The same conundrum applies to things such as apples when we transition from counting them to determining their properties (size, color, etc.). The problem is to find a way to correctly relate the count of apples to some property of apples such as their weight. We can select similarly appearing varieties, shapes, and sizes with reasonable utility, but we cannot determine their differences with complete accuracy. Nor can we generalize about the weight of apples. The count may not suffer from numerical inaccuracy, but we lack a measure; we lack an understanding of some intrinsic property.

Rasch ([1960] 1980) provided a useful model for producing sample-free and item-free measures from counts scored correct or incorrect, applicable to any dichotomous condition or question. Ratings can be addressed likewise, as has been explained in other publications. Since the Rasch model resolves problems in producing sample-free items, and item-free measures from counts, what remains on the agenda is the problem of units.

When dealing with things that are objects or physical entities, it is the individual thing, assumed equivalent and exchangeable, that is the unit of discourse. In this case, the thing usually defines itself, however arbitrary that condition may be. As long as we deal with an object-like thing, the unit derives its properties from the thing itself (always taking into account errors from producing the thing itself). Using an arbitrary standard is not the issue; we have established arbitrary but useful units such as the foot and the meter.

The basic criterion for any unit apart from its arbitrary standard is its pragmatic utility and its portability so that the unit can be produced irrespective of time, place, and person. The unit must have this practicality, which means it must be one that can be envisioned and produced by knowing its specifications. A meter can be copied from the master just as many other units are simply copies of an idealized entity.

The problem is to devise generalized and practical units for measuring things that are not as easily differentiated as physical objects, such as attributes, conditions, and abstractions that will be serviceable to anyone desiring their use. The Système International des Unités or SI units provide this service in science. The SI includes the meter as unit of length; kilogram as unit of mass; second as unit of time; ampere as unit of electric current; Kelvin as unit of temperature; and candela as unit of luminosity. These explicit units are supplemented by others, including a variety of derived units. Some units are single as indicated, while others provide compound units describing physical relationships. A triple one includes the Newton second per square meter—force by time divided by area. Furthermore, these units are under constant assessment.

The US dollar is a simple unit—a measure. Each dollar is equal and exchangeable. It is a measure albeit an arbitrarily devised thing of no intrinsic use, merely paper. However, it can be manipulated mathematically and applied to determine the cost or sale of things, and the worth or loss of objects and entities, some physical and others fictional. Consider one story in the history of the arbitrary foot:

> ... bid sixteen men to stop, tall ones and short ones, as they come out the church door Sunday with 'their left feet one behind the other.' The sum is the rod and the average is 'the right and lawful foot.' (Klein 1974, 67)

Note the random aspects conveyed in this process described by Master Köebel in his text on surveying practices producing the rod and foot. An etching depicting the scene also shows three observers overseeing the process.

More units are required, especially in the social sciences, but this requires theoretical-experimental skill to assure that the unit achieves utility. It is important to note a subtle attribute of the "foot." The foot is much less a physical object than commonly assumed but more a product. The above vignette illustrates "production" of the arbitrary foot. Emphasis should be given to the practical determination of the foot and not be distracted by the object designated "foot." The physical unit is a manifestation and realization of a logico-mathematical process. In the vignette above, it was a simple average from a presumed random sample of church attendees. While historically simple, this example paves the way to more sophisticated expressions of units. The essential property of the unit supports the attribute and vice versa—this is the acid test. A unit must be useful and practical with the capacity to be duplicated across time and space in such a way as to prove its value as a product. The social sciences must seek production of such units or there will be a serious limitation to making progress in social science measurement. Rasch has provided a model for producing measures from counts. The task ahead is to produce practical and useful units that satisfy the conditions for units. Units are especially needed in psychology, sociology, economics, education, and the other social sciences that have not yet established practical and useful units that can be employed in a generic fashion.

The properties of such units should include practicality, theoretical-experimental corroboration, and specificity within a standard error. The road to these units will parallel that of physical ones, with a long road of approximations until a consensus produces practicality and utility. Even at that stage, further refinement and quality control will be required.

There are three essential attributes to measuring. Rasch (1968, 1977, 2001) addressed them in various papers, and Stone and Stenner (2014) further explicated them by focusing on his ashtray example:

1. Comparison is primary; to compare is to distinguish: a = b; a < b; a > b.
2. Order follows: If a < b, b < c, then a < c. Transitivity results, but a test for variable monotonicity is required to establish valid order.
3. Equal Differences require a Standard Unit. A Standard Unit is established either by a personage with power (king, pope), or by agreement arising from data and consensus (science).

The process is developmental; nothing in measuring arises full-blown. Progress occurs by steps of continued understanding resulting from intuition, reason, and improved instrumentation. We designate this encompassing process *measuring mechanisms* (Stenner, Stone, and Burdick 2009).

Two examples illustrate the points we have made. The Mohs scale of hardness is based on a scratch test mechanism. Ten key values range from talc (#1) to diamond (#10). Comparison and order are satisfied. Equal differences are not satisfied inasmuch as the difference from seven to eight is not the same as the difference of three to four. Unequal differences occur across all the scale values. Such a scale is termed ordinal by Stevens and others embracing a level of measurement schema. While the Mohs test is a well-recognized ordinal scale, the Vickers test is a different matter.

The Vickers scale is used in engineering and metallurgy and operates via two different measuring mechanisms: indentation hardness and rebound hardness. The former is determined from a microscopic device equipped with a micrometer for measuring permanent deformation of the material tested. The indentation made from an experimental indenter is carefully measured, resulting in a linear numerical value that is amenable to mathematical operations.

Rebound hardness is measured from the upward "bounce" of a carefully engineered hammer descending from a fixed height (see Stone and Stenner's 2014 explication of Rasch's ashtray dropping experiment— there is similarity, but Rasch employs a purely qualitative approach to make his case). The Vickers experiment uses a scleroscope to provide a precise linear measure of rebound height. Two other related scales for measuring rebound hardness not discussed here are the Leeb rebound hardness test and the Bennett hardness scale.

It would be interesting to apply and to compare the "hardness" of the ten key minerals on Mohs scale to an exact measure of rebound from applying the Vickers measuring mechanism. One might expect this "predicted" scale to reflect an outcome similar to a Winsteps map of items. From a knowledge of minerals, one might expect talc (#1) and gypsum (#2) to be calibrated close to one another because both can be scratched with the fingernail. Topaz (#8), corundum (#9), and diamond (#10) might be somewhat close together at the other end of the scale inasmuch as these three represent gemstones. The remaining minerals might be scattered within the range, close to one another, or in different clusters. The map would indicate "measured" differences in hardness calibrations for the key Mohs indicators whereby some of the ten might be widely separated and others closely placed or similar. Extensive lists of gemstone hardness indicate that most gemstones have Mohs values of seven and higher.

CIDRA Precision Services LLC (2012) published data on mineral hardness showing the relationship of the Mohs scale to the Vickers. A plot of their data is given in figure 1 modeled by a power curve with an $R^2 = 0.9857$. The first four values of the Mohs scale are found between zero and five hundred on the Vickers scale, while the last three values on the Mohs scale show adjacent differences of about five hundred.

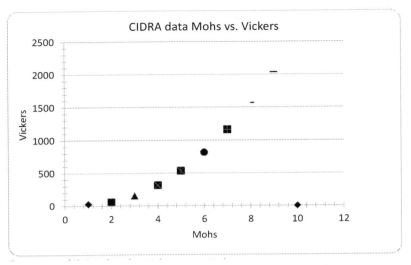

Figure 30.1 CIDRA data for Mohs versus Vickers

Oppenheimer (1956) emphasized that "Analogy is an instrument in science." He identified analogies as vital and indispensable to conducting science. Hardness measured by the Vickers test for indentation or upward bounce progresses analogously beyond (1) comparison and (2) order by providing measures of hardness whereby (3) unit differences are expressed on an equal interval linear scale.

Some mechanisms yield relations that are merely ordinal, while others yield quantitative relations (i.e., homogeneous differences up and down the scale). One explanation for repeated failure to engineer a mechanism sensitive to variation in homogeneous differences for an attribute is that the attribute is ordinal. However, if the history of science is any guide, it may take a century or two before we confidently conclude that an attribute is in some fundamental sense merely ordinal. The odds are that we can achieve measuring properties from ordinal data as Rasch did from his ashtray example, and as CIDRA did with data for Mohs versus Vickers.

These steps can be summarized as follows:

From Ordinality to Quantity

Qualitative	Quantitative
Comparison [more, less, equal]	**Difference** [with a constant unit]
Order (transitivity)	**Trade-off** [demonstrable when quantativity is possible]
Marks and signs of quantativity	**Useful Logit Domain [range]** (agents, objects, reactions}
Successful predictions from differences	

Comparison [equal, more, less] produces order, and qualitative data enhanced by their differences expressed in a constant unit produces quantitative information.

Box 31: Theory and Data

Examining data may lead to conjecture and hypothesis, but it is the formulation of a theory that drives empirical investigations. We explicate the roles of theory and data emphasizing that substantive theory trumps data. In support of this position, we offer a number of examples and testimony.

Our first example comes from the work of Edmund Halley, namesake of the famous comet. Seneca (1928) wrote in one of his essays more than two thousand years ago,

Someday there will arise a man who will demonstrate in what regions of the heavens the comets take their way; why they journey so far from the other planets; what their size, their nature.

That person was to be Edmund Halley (1656–1742). Although he did not succeed in solving all the matters Seneca called for, neither did he accomplish this work alone. Halley was the second astronomer royal at Greenwich, England, a friend and colleague of Newton. Meeting Newton in 1684, Halley would later help Newton with producing *Principia*, especially the section on comets, and he was responsible for seeing to its publication, as Newton was always reluctant to publish his findings.

Halley had already observed numerous comets in 1680 and 1682 and worked out their paths in accordance with Newton's principles of gravitation. He also calculated twenty-four orbits for additional comets that were reported in his *Synopsis of Cometary Astronomy*.

The well-known and frequently reported celestial object that came to be known as Halley's Comet can be traced in Chinese annals back to 1059

BCE. It may have even occurred in a report dated 2467 BCE. Tycho Brahe, the great celestial observer, recorded that this comet was seen in 1607 and earlier in 1531. Brahe speculated that comets follow a path through the solar system from far beyond the moon.

Every return of Halley's Comet since 240 BCE has been observed and recorded. The comet was pictured in the *Bayeux Tapestry* of 1066. Giotto di Dondone recorded its appearance in his painting, *The Adoration of the Magi*. Halley himself observed the comet in 1882 and agreed with Tycho Brahe that it was identical to the comets recorded in 1607 and 1531. Basing his calculations on Newton's theory of gravitation, he predicted a return of this particular comet in 1758–9. Although Halley had correctly conjectured the orbit, the final calculations determining its trajectory were later completed by Alesis Clairaut (1723–1765). The prediction was off by only a month and a day. The amount of error associated with this prediction is extraordinarily small if we use thirty-one days error in covering a span of sixteen years (about 31 / 5840 = 0.005).

The comet's arrival subsequent to Halley's prediction was first recorded by George Palitzsch (1723–1788) of Saxony. Occurring on Christmas day of that year, it made this appearance sixteen years after Halley had died. The Venus transits Halley had deduced using a similar procedure were also validated from observations made two years later. Today we recognize Halley's Comet and its seventy-six-year period. Its last appearance occurred February 9, 1986. Comets are now thought to originate from the Oort Cloud, at a distance of about one light-year. Halley's Comet is deemed an ellipse going around the sun and around the planet Neptune. Its next appearance will be in 2061–2.

What we learn from this interesting story is how the careful collection of historical observations (Tycho Brahe et al.) collated by Halley, together with his application of Newton's theory on gravitation, were combined to provide an estimate of how this comet made its past appearances. Most important of all was the prediction of its next occurrence. Subsequent confirmation after his death makes Halley's prediction a superb example of the connection between theory, observation, and a mathematical model. While many prognosticators have given us bizarre predictions of the future, Halley's forecast stands out as unique and valid. Many of Halley's other predictions have been subsequently confirmed as well.

Urbain Leverrier (1811–1877) in France and John Crouch Adams (1819–1892) in England sought to account for an anomaly in Mercury's orbit. In 1845, Leverrier assumed a planet might exert a gravitational pull on the orbit of Uranus. He calculated the position and time it should be seen, and on September 23, 1846, it was observed at the Berlin Observatory by Johann Galle.

The fame of Leverrier brought him a letter from Lescarbault indicating that he had observed a new planet passing in front of the sun. Leverrier subsequently visited Lescarbault. Based upon this meeting and only this one sighting, Leverrier computed the planet's orbit. Subsequent years proved this prediction false. Not all predictions are valid. Recognizing failure is as important, if not more so, than celebrating success.

Johann Daniel Titius (1729–1796), and Johann Elert Bode (1747–1826) are famous for another prediction. Consider the numbers 0, 3, 6, 12, 24, and 48, then add 4 to each, producing 4, 7, 10, 16, and so on, and you end up with approximations of the relative distances of planets in our solar system from the sun, with Earth being 10. The Titius-Bode Law is empirically based with no theoretical basis beyond the numerical process. Interestingly, only Neptune is noticeably discrepant from this pattern.

	Titius-Bode	Distance		Titius-Bode	Distance
			Titius-Bode	1	
Mercury	0.4	0.39	Distance	0.992525	1
Venus	0.7	0.72			
Earth	1.0	1.00			
Mars	1.6	1.52			
Ceres	2.8	2.77			
Jupiter	5.2	5.20			
Saturn	10.0	9.54			
Uranus	19.6	19.19			
Neptune	38.8	30.07			

A contrived law from observed data may appear striking, but lacking a theoretical basis, there is no convincing theory to support the contrived data. Offering only a numerical argument based on a set of data is not sufficient. The Titius-Bode is only an interesting numerical curiosity.

By contrast, a unique example comes from the work of Dimitri Mendeléyev (1834–1907). His conjecture in 1869 predicted that "the elements arranged according to the magnitude of atomic weights show a periodic change of properties" (Brock 1992). While other investigators were also arriving at similar conjectures, we usually associate

Mendeléyev's name to the origin of the periodic table. The value for his discovery of the periodic law was shown by his eventual finding that the properties of seventeen additional elements could be moved into new and more accurate positions. His theory also showed there were errors in the previously accepted atomic weights, small for some, larger for others. Intriguingly, Mendeléyev was able to predict the existence of many of the properties for yet undiscovered elements. His predicted elements eka-boron, eka aluminum, and eka-silicon [eka is Sanskrit for "one"] became what was later discovered to be *scandium, gallium,* and *germanium*. The following table illustrates the correspondence between his theory values and the later experimental data for the atomic weight of these three elements.

Element	Predicted Atomic weight	Experimental Determination
Eka-aluminum	68	69.9
Eka-boron	44	43.79
Eka-silicon	72	72.3

These and other successes gave strong affirmation to the predictive power of the periodic law. Not all of Mendeléyev's predictions and recommended adjustments were correct, but that should not detract us from appreciating the predictive power implied by his approach. Chemistry subsequently experienced a more useful dialectic between theory emanating from the discovery of periodicity and better experimental work. This served as a basis for discoveries with each new aspect in the dialectic enhancing another.

Bohr later conjectured that element 72 should be close to the expected location of zirconium according to its position in the periodic system. Further experimental work by de Hevesy and Coster in 1922 found the missing element, which they named Hafnium, located where it was predicted to be found. The theory emanating from the periodic law has been found useful in yielding practical results. It simplified many aspects of chemistry that were otherwise confusing.

A further example concerns the prediction of an unseen companion of the very bright star in Cannes Major named Sirius. This description is given by Burnham in his *Celestial Handbook,* volume I:

> In the years between 1834 and 1844 the astronomer and mathematician F. W. Bessel found that Sirius had wavy irregularities in its motion through space, and came to the conclusion that the star had an invisible companion revolving about it in a period of about 50 years. The theoretical orbit of this unseen body was actually calculated in 1851 by C. H. F. Peters, but the expected companion [Sirius B] persistently refused to show itself, despite the careful searches made by many experienced observers. Then, in January 1862, the prediction was fulfilled by the discovery of the companion near its expected place by Alvan G. Clark with an 18 1/2-inch refracting telescope, then the largest refractor in the world. (394–95)

... From this result it is an obvious step to the final, most amazing characteristic of the 'Sirius B' stars, now called 'white dwarfs.' The typical white dwarf must have an incredibly high density. Sirius B has a density of about 90,000 times that of the sun, or about 125,000 times the density of water. A cubic inch of this star material weighs about 2 1/4 tons. (399)

Burnham's story is another example of the power contained in the careful delineation and application of a valid theoretical model. We also note that confirmation by data may require a substantial investment of time and instrumentation in order to validate a theory. Confirmation of Einstein's earliest predictions, for example, took about twenty years and was achieved when the astronomer Eddington observed light "bending" during the solar eclipse of May 29, 1919.

Persons looking for quick solutions and confirmations will be frustrated by the time required for some experiments to be conceptualized, conducted, and validated. Sometimes it is lack of sensitive instrumentation that frustrates experiments. Only when instrumentation is adequate to the task can some theories be realized. These examples serve as striking illustrations of the power emanating from useful theories when supported by experimental findings.

The integration and inter-relationship of theory and data has long been the focus of scientists. This history has been recorded by philosophers of science, but the remarks of the scientists themselves about the dialogue between theory and data is revealing. While Newton determined the properties of gravitation that constituted the "what" part of gravitation, he was reluctant to state the "why" with certainty. He wrote in *Principia*:

> I have not yet been able to deduce from phenomena the reason of these properties of gravitation, and I do not invent hypotheses [Hypotheses non fingo]. For anything which cannot be deduced from phenomena should be called a hypothesis. (795)

Newton clearly distinguished between that which is known and that which is surmised. The latter was designated a hypothesis. Newton's quote using "hypothesis" should be distinguished from contemporary usage. He was speaking of hypothetical speculation after the fact. Correspondence between reason based on theory and that based on phenomena must be clearly established, or we have only conjectures and "hypotheses."

Comte's attempt to build a comprehensive positive philosophy was to be established upon the foundations of science. Although not all Comte's ideas satisfied his own criteria, he did provide some useful remarks, including the following:

... theories consist essentially of laws and not of facts, although the latter are indispensable for their justification and confirmation. Thus no isolate fact, no matter of what sort, can really be incorporated into science before it is at least correctly connected with some other conception through the aid of a rational hypothesis. (6)

> There is close similarity between Comte and Newton with respect to their positions on this matter. Comte stressed the interdependence of theory and data. Theory provides generalizations, which must correspond to facts. Furthermore, facts, especially isolated ones, offer no true understanding of reality.

Writing more than one hundred years ago, the astronomer Berry described the relationship of theory and data:

> Facts are obtained by observation or experiment; a hypothesis from provisional theory is devised to account for them; from this theory are obtained, if possible by a rigorous process of deductive reasoning, certain consequences capable of being compared with actual facts, and the comparison is then made. (245)

Confirmation of the remarks of Newton and Comte is provided by Berry showing that science is an amalgam of theory and facts cemented together by careful reasoning. The role of the hypothesis is to provide a "provisional theory." Joad (1932) asserted that "Science is based upon observation of the physical world, and reaches its conclusions by reasoning about what it has observed" (94).

Poincare (1904) makes an additional claim for integrating theory and data:

> It is the mathematical physics of our fathers which has familiarized us little by little with these divers principles; which has taught us to recognize them under the different vestments in which they disguise themselves. One has to compare them [these principles] to the data of experience, to find how it was necessary to modify their enunciation so as to adapt them to these data; and by these processes they have been enlarged and consolidated. (604)

A stronger caveat comes from the astronomer Eddington (1921):

> It is also a good rule not to put too much confidence in the observational results that are put forward until they are confirmed by theory. (211)

"Confirmed by theory" is an important phrase. Eddington argues that facts alone are never the answer, and that data must be affirmed by a valid theory. This approach is the opposite of what is typically proposed—theory produced by examining data.

Parsons, late professor of sociology at Harvard, writes, "All empirically verifiable knowledge—even the common-sense knowledge of everyday life—involves implicitly, if not explicitly, systematic theory" (10).

Theory—that is, systematic theory—is the touchstone of understanding according to Eddington and Parsons.

The physicist Braithwaite confirms their comments:

> The function of a science, in this sense of the word, is to establish general laws covering the behavior of the empirical events on objects with which the science in question is concerned, and thereby to enable us to connect together our knowledge of the separately known events, and to make reliable predictions of events as yet

unknown. ... The fundamental concept for science is thus that of scientific law, and the fundamental aim of a science is the establishment of such laws. (1–2)

Herbert Simon (1951) affirms this position.

What distinguishes science from every other form of human intellectual activity is that it disciplines speculation with facts. Theory and data are the two blades of the scissors. (110)

These examples and quotations lend support to the position that theory must drive investigations while data plays only a supportive role.

Box 32: Why We Write the Way We Do

Although we are certainly concerned with reliability, validity, and other psychometric issues (see box 32, "Perspectives on Reliability and Validity"), our emphasis is upon actions we deem more appropriate and essential (see box 3, "From Semantics to Process"). The business of measuring is one of constructing, fabricating, and engineering. These terms and the actions they provoke are what we deem most important. We are especially comfortable using recipe, prescription, and engineering blueprints. These words evoke the theory, constructive, and causal approach we deem essential to making measures.

The construction and production of measuring instruments has much to gain from shifting words and content from the psychometric realm to those of engineering. We are in the business of building measuring instruments. We find support from Simon's argument (1969), "It has been the task of the science disciplines to teach about natural things: how they are and how they work. It has been the task of engineering schools to teach about artificial things: how to make artifacts that have desired properties and how to design." (p. 56).

Consequently, we find that the words we embrace align with Simon's position as richer and more powerful tools by which to guide the construction and production of measures than those supplied by reiterating overused psychometric jargon.

We design the process by which measures are to be made. We construct the process by which instrumentation and measures are to be produced, manufactured, or fabricated. The production process is one that requires quality control to continually monitor the production process.

Changes are made accordingly. Fix and improve what needs to be changed. Continuously monitor the results.

Guttman writes on this point suggesting (1991), "A basic reason for the persistence of informality is the developmental nature of science. Perception of new things is of necessity partly in non-technical 'primitive' terms. ... "A first lesson in learning how to do science may well be on the respective roles of informal and formal thinking and on the need for a strategy for optimal transformation of informality into formality. ... "At least three levels are employed simultaneously: (1) mother tongue (ordinary speech), (2) technical terminology, and (3) formulae (analytic symbolization) (p. 1).

Our perspective on Guttmann's three levels of discourse constitutes (1) a substantive theory and causal model, (2) appropriate instrumentation, and (3) a specification equation (see "Measurement Mechanism" and box 24).

Box 33: Validity Is Theoretical Equivalence

Forty years ago in the *Australian Journal of Psychology,* Lumsden and Ross (1973) proposed "Validity as theoretical equivalence." We affirm this proposition still further by proffering "Validity *is* theoretical equivalence." Lumsden and Ross defined validity to be what a test measures successfully. This affirmation does not suggest that all tests, sui generis, are valid. Determining what systematic variance an instrument detects is crucial to establishing "validity." Furthermore, a theoretical basis contributes the essential substantive element for constructing tests demonstrating validity, with prediction based on substantive theory being the hallmark of success.

Guttman (1971) advocated abandonment of the term *measurement* as a concept in favor of predictive measures, that is, determining the interaction between a conceptually realized measure and some empirical application, what he called "structural theory." Rather than relying upon definition(s), Guttman said science should establish relationships employing prediction as the critical determinant. Relying on definition(s) is passive, while prediction is action oriented and constructive.

Two examples come to mind: The first is "distance equals rate times time," D = RT. Having determined this relationship by an explicit equation, one can solve (predict) for D, R, or T by algebraically utilizing the other two values. We can predict travel outcomes for time, distance, and rate of speed by this equation. The second is the common thermometer, which relates a graduated column of mercury to environmental states of cold to hot.

Having substantiated this connection, one can measure environmental temperature by means of the thermometer and, more importantly, predict the environmental state by knowledge of the temperature. With this connection validated, knowing the mean temperature of a country in the month of May allows one to determine what clothes to bring and what to leave behind. Successful measuring mechanisms facilitate prediction under intervention, which is the hallmark of success (i.e., validity).

These examples indicate that a successful *measuring mechanism* (Stenner, Stone, and Burdick 2009) fabricates a bridge between theory-calibrated mechanism (thermometer) and some reality-based substance or condition (e.g., environmental temperature). Validity is the predictive success established by a constructed bridge between these two elements, allowing prediction of one condition from knowledge of the other. Validity is theoretical equivalence.

Lumsden and Ross (1973) "argued that "the general steps in the establishment of theoretical equivalence may be summarized as follows:

(a) Selecting a technique based on observed covariation with the operational criterion of a theoretical term.

(b) Removing unwanted systematic variance by the use of available technical and theoretical developments.

(c) Demonstrating that the techniques yield relationships congruent with the theory term equivalent or with some new theory term postulated to account for perturbations or new phenomena discovered with the technique.

(d) Accepting the measure as equivalent to the theoretical term." (193)

What has prevented this perspective on *validity* from being mainstream? Two issues stand out. The first is employing complex theories that are not straightforward, clear, and robust in their conception. Second, failing to proceed beyond data-based instrument calibration or statistical conclusions in order to demonstrate the connection between theory and data.

Guttman (1972) stated,

Physicists formulate hypotheses about relationships among masses and distances and other concepts, and the assignment of numbers they make is on the basis of these hypotheses ... a structural theory has to be employed: one must specify how the observations are made and what regression systems are hypothesized. (341)

He subsequently presented a problem:

Let ϱ_{xy} be the population correlation coefficient for the normal bivariate distribution of x and y. Draw a random sample of n observations, and test the null hypothesis that $\varrho_{xy} = 1$ against the alternative that $\varrho_{xy} < 1$. This is a straightforward problem is statistical inference, yet has evoked some surprising reactions. (344)

This statement in the form of a differences surprises all whose expectation for an experimental outcome is, hopefully, statistically significant from 0.00, rather than present a well-executed substantively-based theory proffered to be at or near r'=1.0 (corrected for artifacts like range restriction and measurement error). Social scientists, as Taagepera (2008) clearly points out, do not think along these lines as do physicists. They see life as unaccountably complex, which is true, but then propose equally complex models impossible to actualize. Consequently, they remain content to agonize over low coefficients that illustrate the "complexity" of life while being unable to understand the "problem" Guttman proposed.

Social scientists reason differently from physicists. Taagepera (2008) makes these comparisons:

1. Social science equations are deemed too complex with too many variables,
2. Social scientists are content to describe, physicists strive to predict,
3. Physicists tend to favor sequential processes; social scientists assume simultaneous impact,
4. Physicists utilize algebraic equations which are commutative and symmetric, and
5. Physicists multiply, social scientists add, employ logarithms.

Lumsden and Ross (1973) furthermore state:

The fundamental validity question for test theory is: what does the test measure? The validity problem is to discover how to characterize the systematic variance of the test in an acceptable way. Only construct validity (APA, 1954; Cronbach and Meehl, 1955) attempts to answer the validity question as we have put it. Construct validity is used when it is desired to infer the degree to which an individual possesses some trait or quality (construct) presumed to be reflected in a test performance (APA, 1954, p. 13). (191)

A useful context for examining theoretical equivalence is the relationship between temperature and thermometers.

(a) Selecting a technique based on observed co variation with the operational criterion of a theoretical term.

(b) Removing unwanted systematic variance using available technical and theoretical developments.

(c) Demonstrating that the techniques yield relationships congruent with the theory term equivalent or with some new theory postulated to account for perturbations or new phenomena discovered with the technique.

(d) Accepting the measure as equivalent to the theoretical term. (193)

Quotes from Lumsden and Ross (1973) indicate they understood the essence of validity and how it is to be demonstrated in practice:

> The validity problem in physics is solved when equivalence is established between a theory term and a measure in the temperature-thermometer sense. A measurement procedure is regarded as valid, if, and only if, its systematic variance can be shown to be equivalent to one, and only one, theoretical term. (192–193)

Lumsden and Ross more than fifty years ago advocated a position like the position we advocate in the chapters and boxes given in *Substantive-Theory and Constructive Measures*.

Thus, an instrument is valid if it transmits variation in the intended attribute (temperature or reading ability) to the measurement outcome (often a count). To give this formulation teeth, "intention" must be represented independent of the attribute label. It is insufficient to assert that this test (e.g., Test of English Reading Ability) is valid because, for example, TERA test data fit a Rasch model and demonstrate group invariance. So, how does intention get specified and tested? During the last one hundred years of testing practice, most answers to this question involved correlations with other tests with similar labels. Some theorists have argued for specification of the intended attribute by linking test performance to micro processes like short-term memory, perceptual speed, and so on (Embretson 2007; Gorin 2007; Pellegrino 1988). Others (Wright and Stone 1983; Stenner, Smith, and Burdick 1983) chose to focus on specifying those features of the measurement mechanism that are critical to transmitting attribute variation to the measurement outcome while recognizing that different task types might employ different mechanisms to transmit variation in the self-same attribute (e.g., NexTemp versus mercury in tube thermometers). Measurement mechanisms are much more accessible to manipulation than are subcomponents of human behavior taken to be the more basic constituents of the target attribute (e.g., phonological facility or short-term memory as microprocesses underlying reading ability). In our view, focus on measurement mechanisms has been more fruitful in explicating what an instrument measures than process approaches. However, we should point out that our focus on measurement mechanisms should not be interpreted as support for nomological networks. We agree with Gorin (2007) that validity should have as its focus how the measurement instrument works. We choose to privilege "mechanism" (features of the instrument) over "processes" (features of the object of measurement).

REFERENCES

Alexander, H. 1947. "The Estimation of Reliability When Several Traits Are Available." *Psychometrika* 12:79–99.

American Educational Research Association, American Psychological Association, and National Council on Measurement in Education. 1999. *Standards for Educational and Psychological Testing.* Washington, DC: American Educational Research Association.

ANSUR: Open Design Lab at PSU. 1988. http://openlab.psu.edu/.

Andrich, D. 1982. "An Index of Person Separation in Latent Trait Theory, the Traditional K.R.20 Index, and the Guttman Scale Response Pattern." *Educational Research and Perspectives* 9:95–104.

———. 1985. "An Elaboration of Guttman Scaling with Rasch Models for Measurement." In *Sociological Methodology,* edited by N. Brandon-Tuman, 30–80. San Francisco: Jossey-Bass.

———. 1988. "Rasch Models for Measurement." *Sage University Paper Series on Quantitative Applications in the Social Sciences,* vol. 07-068. Beverly Hills, CA: Sage Publications.

———. 2002. "Understanding Resistance to the Data-Model Relationship in Rasch's Paradigm: A Reflection for the Next Generation." *Journal of Applied Measurement* 3:325–59.

———. 2004. "Controversy and the Rasch Model: A Characteristic of Incompatible Paradigms?" *Medical Care* 42:1–16.

———. 2010. "Sufficiency and Conditional Estimation of Person Parameters in the Polytomous Rasch Model." *Psychometrika* 75:292–308.

Andrich, D., and I. Styles. 1994. "Psychometric Evidence of Intellectual Growth Spurts in Early Adolescence." *Journal of Early Adolescence* 14 (2): 126–49.

Arnold, D. 1991. *Building in Egypt: Pharaonic Stone Masonry*. Oxford: Oxford University Press.

———. 2003. *The Encyclopaedia of Ancient Egyptian Architecture*. Princeton, NJ: Princeton University Press.

Augustine, A. 1957. *The Confessions of St. Augustine*. New York: Pocket Books.

Baker, D. 1978. *Astronomy*. New York: Henry Holt.

Barkay, G. 1986. "Measurements in the Bible: Evidence at St. Etienne for the Length of the Cubit and Reed." *Biblical Archeological Review* 12 (2): 37.

Barlow, D. H., M. K. Nock, and M. Hersen. 2009. *Single Case Experimental Designs*, 3rd ed. Boston: Pearson.

Barrois, G. A. 1952. "Chronology and Metrology." In *The Interpreter's Bible*. Vol. 1. New York: Abingdon Press.

Bejar, I., R. Lawless, M. Morley, M. Wagner, R. Bennett, and J. Revuelta. 2003. "A Feasibility Study of On-the-Fly Item Generation in Adaptive Testing." *Journal of Technology, Learning, and Assessment* 2:1–29. http://ejournals.bc.edu/ojs/index.php/jtla/article/view/1663.

Berriman, A. D. 1953. *Historical Metrology*. London: Dent.

Berry, A. 1961. *A Short History of Astronomy*. New York: Dover.

Binet, A. (1908) 1916. *The Development of Intelligence in the Child*. Paris: L'Année Psychologie. Reprint, Baltimore: Williams & Wilkins. Citations refer to the Williams & Wilkins edition.

BIPM. n.d. "Unit of Time (Second)." *SI Brochure*. Accessed December 22, 2013. BIPM.org.

Bollen, K. A., and R. Lennox. 1991. "Conventional Wisdom on Measurement: A Structural Equation Perspective." *Psychological Bulletin* 100:305–14.

Borsboom, D. 2005. *Measuring the Mind*. Cambridge: Cambridge University Press.

Braithwaite, R. (1953) 1956. *Scientific Explanation*. New York: Torchbooks. Reprint, Cambridge: Cambridge University Press. Citations refer to the Cambridge University Press edition.

Bridgman, P. 1928. *The Logic of Modern Physics*. New York: Macmillan.

Brock, W. 1992. *The Norton History of Chemistry*. New York: Norton.

Brody, H. 1993. *Philosophy behind Physics: The Collected Papers of Harold Brody*. Edited by L. De LaPena and P Hodgson. Berlin: Springer-Verlag.

Brody, T. 1993. *The Theory behind Physics*. New York: Springer-Verlag.

Brogdin, H. E. 1977. "The Rasch Model, the Law of Comparative Judgment and Additive Conjoint Measurement." *Psychometrika* 42:631–34.

Bunge, M. 2004. "How Does It Work? The Search for Explaining Mechanisms." *Philosophy of Social Science* 34:182–210.

Burdick, D. S., M. H. Stone, and A. J. Stenner. 2006. "The Combined Gas Law and a Rasch Reading Law." *Rasch Measurement Transactions* 20 (2): 1059–60.

Burnham, R. 1978. *Celestial Handbook*. New York: Dover.

Burt, C. 1922. *Mental and Scholastic Tests*. London: P. S. King.

Buttrick, G. A. 1952. *The Interpreter's Bible*. Vol. 1. New York: Abingdon Press.

Campbell, D., and J. Stanley. 1963. *Experimental and Quasi-Experimental Designs for Research*. Chicago: Rand McNally.

Campbell, N. (1921) 1953. *What Is Science?* London: Methuen. Reprint, New York: Dover. Citations refer to the Methuen edition.

Carnap, R. 1966. *Philosophical Foundations of Physics: An Introduction to the Philosophy of Science.* New York: Basic Books.

Cartwright, N. 1983. *How the Laws of Physics Lie.* New York: Oxford Press.

———. 2001. "Modularity: It Can—and Generally Does—Fail." In *Stochastic Causality*, edited by Maria Carla Galavotti, Patrick Suppes, and Domenico Costantini, 65–84. Stanford: CSLI Publications.

———. 2002. "Against Modularity, the Causal Markov Condition, and Any Link between the Two: Comments on Hausman and Woodward." *British Journal for the Philosophy of Science* 53:411–53.

———. 2003. "Two Theorems on Invariance and Causality." *Philosophy of Science* 70:203–24.

Cellini, B. 1906. *Autobiography: The Life of Benvenuto Cellini.* Translated by J. Symonds. New York: P. F. Collier & Son.

Chang, H. (2004) 2007. *Inventing Temperature: Measurement and Scientific Progress.* New York: Oxford University Press. Reprint, London: Cambridge University Press. Citations refer to the Oxford University Press edition.

Choppin, B. 1968. "Item Banking Using Sample-Free Calibration." *Nature* 219 (5156): 870–72.

———. 1985. "A Fully Conditional Estimation Procedure for Rasch Model Parameters." *Evaluation in Education* 9:29–42.

Collingwood, R. 1940. *An Essay on Metaphysics.* Oxford: Clarendon Press.

Comte, A. 1864. *Cours de philosophie positive.* Paris: Bailliere.

Conant, J. 1947. *On Understanding Science.* New York: Mentor.

Connolly, A., W. Nachtman, and E. Pritchett. 1988. *Manual for the Key-Math Diagnostic Arithmetic Test.* Rev. ed. Circle Pines, MN: American Guidance Service.

Cook, T., and D. Campbell. 1979. *Quasi-Experimentation: Design and Analysis Issues for Field Settings.* Boston: Houghton Mifflin.

Coombs, C. H., R. M. Dawes, and A. Tversky. 1970. *Mathematical Psychology: An Elementary Introduction.* Englewood Cliffs, NJ: Prentice Hall.

Cornford, F. M. 1957. *Plato's Theory of Knowledge: Theaetetus and Sophist.* New York: Liberal Arts Press.

Courtis, S. 1932. *The Measurement of Growth.* Ann Arbor, MI: Brumfield and Brumfield.

Cronbach, L. 1970. Review of *On the Theory of Achievement Test Items,* by J. Bormuth. *Psychometrika* 35:509–11.

Cronbach, L., and P. Meehl. 1955. "Construct Validity in Psychological Tests." *Psychological Bulletin* 52:281–302.

Dantzig, T. 1954. *Number: The Language of Science.* New York: Free Press.

David, A. B. 1969. "Ha-midda ha-Yerushalmit." *Israel Exploration Journal* 19:159–69.

Dawson-Tunik, T., M. Commons, M. Wilson, and K. Fischer. 2005. "The Shape of Development." *European Journal of Developmental Psychology* 2 (2): 163–95.

Dillman, D. 2000. *Mail and Internet Surveys.* New York: John Wiley and Sons.

Dornič, S. 1967. "Subjective Distance and Emotional Involvement: A Verification of the Exponent Invariance." *Reports from the Psychological Laboratories,* University of Stockholm, no. 237.

Dowe, P. 2000. *Physical Causation.* Cambridge: Cambridge University Press.

Dreyer, J. L. E. 1953. *A History of Astronomy from Thales to Kepler.* New York: Dover.

Easton, M. G. 1897. *Illustrated Bible Dictionary*. Knoxville, TN: Thomas Nelson.

Eddington, A. 1935. *New Pathways in Science*. Cambridge: Cambridge University Press.

Edwards, J. R., and R. P. Bagozzi. 2000. "On the Nature and Direction of Relationships between Constructs and Measures." *Psychological Methods* 5:155–74.

Einstein, A. 1902. "Kinetische Theorie des Wärmegleichgewichtes und des zweiten Hauptsatz der Thermodynamik." *Annalen der Physik* 9:417–33.

Ekman, G., and O. Bratfisch. 1965. "Subjective Distance and Emotional Involvement: A Psychological Mechanism." *Acta Psychologica* 24:446–53.

Ellis, J. L., and A. L. Van den Wollenberg. 1993. "Local Homogeneity in Latent Trait Models: A Characterization of the Homogeneous Monotone IRT Model." *Psychometrika* 58:417–29.

Elster, J. 1989. *Nuts and Bolts for the Social Sciences*. Cambridge: Cambridge University Press.

Embretson, S. E. 2006. "The Continued Search for Nonarbitrary Metrics in Psychology." *American Psychologist* 61 (1): 50–55.

Englehard, G. 2008. "Historical Perspectives on Invariant Measurement: Guttman, Rasch, and Mokken." *Measurement: Interdisciplinary Research and Perspectives* 6:155–89.

Feller, W. 1940. "On the Logistic Law of Growth and Its Empirical Verification in Biology." *Acta Biotheoretica* 5:51–65.

Feynman, R. 1965. *The Character of Physical Law*. Cambridge, MA: MIT Press.

Fink, A. 1995. *The Survey Kit 2: How to Ask Survey Questions*. Thousand Oaks, CA: SAGE.

Fisher, R. 1922. *On the Mathematical Foundations of Theoretical Statistics*. London: Philosophical Transactions of the Royal Society.

———. 1951. *The Design of Experiments*. New York: Hafner.

———. 1958. *Statistical Methods for Research Workers*. New York: Hafner.

Fisher, W., Jr. 1997. "Physical Disability Construct Convergence across Instruments: Towards a Universal Metric." *Journal of Outcome Measurement* 1:87–113.

———. 1999. "Foundations for Health Status Metrology: The Stability of MOS SF-36 PF-10 Calibrations across Samples." *Journal of the Louisiana State Medical Society* 151:566–78.

———. 2000a. "Objectivity in Psychosocial Measurement: What, Why, How." *Journal of Outcome Measurement* 4:527–63.

———. 2000b. "Rasch Measurement as the Definition of Scientific Agency." *Rasch Measurement Transactions* 14:761.

———. 2005. "Daredevil Barnstorming to the Tipping Point: New Aspirations for the Human Sciences." *Journal of Applied Measurement* 6:173–79.

———. 2009. "Invariance and Traceability for Measures of Human, Social, and Natural Capital: Theory and Application." *Measurement* 42:1278–87.

———. 2011. "Bringing Human, Social, and Natural Capital to Life: Practical Consequences and Opportunities." In *Advances in Rasch Measurement*, edited by N. Brown, B. Duckor, K. Draney, and M. Wilson, 2:1–27. Maple Grove, MN: JAM Press.

———. 2012a. *NIST Critical National Need Idea White Paper: Metrological Infrastructure for Human, Social, and Natural Capital*. Washington, DC: National Institute for Standards and Technology. http://www.nist.gov/tip/wp/pswp/upload/202_metrological_infrastructure_for_human_social_natural.pdf.

———. 2012b. "What the World Needs Now: A Bold Plan for New Standards." *Standards Engineering* 64.

Fisher, W., Jr., R. Harvey, and K. Kilgore. 1995. "New Developments in Functional Assessment: Probabilistic Models for Gold Standards." *NeuroRehabilitation* 5:3–25.

Fisher, W., Jr., and A. J. Stenner. 2012. "Metrology for the Social, Behavioral, and Economic Sciences." Social, Behavioral, and Economic Sciences White Paper Series. Washington, DC: National Science Foundation. http://www.nsf.gov/sbe/sbe_2020/submission_detail.cfm?upld_id=36.

Freedman, D. 1997. "From Association to Causation via Regression." In *Causality in Crisis? Statistical Methods and the Search for Causal Knowledge in the Social Sciences*, edited by V. McKim and S. Turner, 113–61. South Bend, IN: University of Notre Dame Press.

Freedman, D. N. 2000. *Eerdman's Dictionary of the Bible*. Grand Rapids, MI: Eerdmans.

Galton, F. 1863. *Meteorgraphia: Methods of Mapping the Weather*. London: Macmillan.

———. 1869. *Hereditary Genius: An Inquiry into Its Laws and Consequences*. London: Macmillan.

———. 1888. "Co-relations and Their Measurement Chiefly from Anthropometric Data." *Proceedings of the Royal Society of London* 45:135–45.

———. 1889. *Natural Inheritance*. London: Macmillan.

———. 1890. "Kinship and Correlation." *North American Review* 150:419–31.

———. 1892. *Fingerprints*. London: Macmillan.

Gasking, D. 1955. "Causation and Recipe." *Mind* 64:479–87.

Gibbs, J. W. 1902. *Elementary Principles in Statistical Mechanics*. New Haven, CT: Yale University Press.

Gillings, R. J. 1972. *Mathematics in the Time of the Pharaohs*. Cambridge, MA: MIT Press.

Gnanadesikan, R. 1977. *Methods for Statistical Data Analysis of Multivariate Observations*. New York: Wiley.

Gorin, J. S. 2005. "Reading Comprehension Questions: The Feasibility of Verbal Item Generation." *Journal of Educational Measurement* 42 (4): 351–73.

Gould, P., and R. White. 1974. *Mental Maps*. New York: Penguin.

Grandjean, E. 1989. *Fitting the Task to the Man*. New York: Taylor & Francis.

Grice, J. W. 2011. *Observation Oriented Modelling*. New York: Elsevier.

Guilford, J. 1954. *Psychometric Methods*. New York: McGraw-Hill.

Guttman, L. 1944. "A Basis for Scaling Qualitative Data." *American Sociological Review* 9:139–50.

———. 1969. "Integration of Test Design and Analysis." In *Proceedings of the 1969 Invitational Conference on Testing Problems*. Princeton, NJ: Educational Testing Service.

———. 1971. "Measurement as Structural Theory." *Psychometrika* 36 (4): 329–47.

Haavelmo, T. 1944. "The Probability Approach in Econometrics." Supplement, *Econometrica* 12.

Hall, N. 2000. "Causation and the Price of Transitivity." *Journal of Philosophy* 97:198–222.

Halpern, J., and J. Pearl. 2005. "Causes and Explanations: A Structural Model Approach." Pt. 1, "Causes," and pt. 2, "Explanations." *British Journal for the Philosophy of Science* 56:843–87, 889–911.

Hambleton, R., H. Swaminathan, and L. Rogers. 1991. *Fundamentals of Item Response Theory.* Newbury Park, CA: Sage Publications.

Hanlon, S. T. 2013. "The Relationship between Deliberate Practice and Reading Ability." PhD diss. The University of Chicago. ProQuest (AAT 3562741).

Hardy, G. H. 1908. *Course in Pure Mathematics.* London: Cambridge University Press.

Hausman, D. 1986. "Causation and Experimentation." *American Philosophical Quarterly* 23:143–54.

———. 1998. *Causal Asymmetries.* Cambridge: Cambridge University Press.

Hedström, P. 2005. *Dissecting the Social: On the Principles of Analytical Sociology.* Cambridge: Cambridge University Press.

Herodotus. 1954. *The Histories.* Translated by Aubrey De Sélincourt with notes by John Marincola. London: Penguin Books.

Hiddleston, E. 2005. Review of *Making Things Happen*, by James Woodward. *Philosophical Review* 114:545–47.

Hitchcock, C. 2001a. "The Intransitivity of Causation Revealed in Equations and Graphs." *Journal of Philosophy* 98:273–99.

———. 2001b. "A Tale of Two Effects." *Philosophical Review* 110:361–96

———. 2007. "Prevention, Preemption, and the Principle of Sufficient Reason." *Philosophical Review* 116:495–532.

Hitchcock, C., and J. Woodward. 2003. "Explanatory Generalizations." Pt 2, "Plumbing Explanatory Depth." *Nôus* 37:181–99.

Hitchcock, H. 2007. "What Russell Got Right." In *Causation, Physics, and the Constitution of Reality: Russell's Republic Revisited*, 45–65. Oxford: Oxford University Press.

Holland, P. W. 1986. "Statistics and Causal Inference." *Journal of the American Statistical Association* 81:945–60.

———. 1990. "On the Sampling Theory Foundations of Item Response Theory Models." *Pshychometrika* 55:577–601.

Hoover, K. D. 2004. "Lost Causes." *Journal of the History of Economic Thought* 26:149–64.

Hoyt, C. 1941. "Test Reliability Estimated by Analysis of Variance." *Psychometrika* 6:153–60.

Hume, D. (1817) 1949. *A Treatise of Human Nature*. Reprint, London: Dent and Sons.

Humphreys, P. 2004. *Extending Ourselves: Computation Science, Empiricism, and Scientific Method*. New York: Oxford University Press.

Hunter, J. 1980. "The National System of Scientific Measurement." *Science* 210:869–74.

Irvine, S. H., and P. C. Kyllonen. 2002. *Item Generation for Test Development*. Mahwah, NJ: Lawrence Erlbaum.

Jackson, R. 1939. "Reliability of Mental Tests." *British Journal of Psychology* 29:269–87.

Jasso, G. 1988. "Principles of Theoretical Analysis." *Sociological Theory* 6:1–20.

Joad, C. 1932. *Philosophical Aspects of Modern Science*. London: Tinling.

Johnson, R. C., G. E. McClearn, S. Yuen, C. T. Nagoshi, F. M. Ahern, and R. E. Cole. 1985. "Galton's Data a Century Later." *American Psychologist* 40 (8): 875–92.

Kaplan, A. 1964. *The Conduct of Inquiry*. New York: Thomas Crowell.

Kant, I. (1798) 1917. *Gesammelte Schriften*. Vol. 7. Berlin: Reimer.

Karabatsos, G. 2000. "A Critique of Rasch Residual Fit Statistics." *Journal of Applied Measurement* 1:152–76.

— — —. 2001. "The Rasch Model, Additive Conjoint Measurement, and New Models of Probabilistic Measurement Theory." *Journal of Applied Measurement* 2:389–423.

Karkach, A. 2006. "Trajectories and Models of Individual Growth." *Demographic Research* 15 (12): 347–400. https://doi.org/10.4054/DemRes.2006.15.12

Kaufman, A. S. 2004. *The Temple of Jerusalem.* Jerusalem: Har Year'ah Press.

Kerlinger, F. 1986. *Foundations of Behavioral Research.* New York: Holt, Rinehart and Winston.

Klein, H. A. 1974. *The Science of Measurement.* New York: Dover.

Komlos, J., and J. Baten. 1998. "Height and the Standard of Living." *Journal of Economic History* 58 (3): 866–70.

Krantz, D. H., R. D. Luce, P. Suppes, and A. Tversky. 1971. *Foundations of Measurement.* Vol. 1, *Additive and Polynomial Representations.* New York: Academic Press.

Kuhn, T. S. 1957. *The Copernican Revolution.* Cambridge, MA: Harvard University Press.

— — —. 1961. "The Function of Measurement in Modern Physical Science." *Isis* 52 (168): 161–93.

— — —. 2008b. *"The Structure of Scientific Revolutions.* Chicago: University of Chicago Press.

Kyngdon, A. 2008a. "Conjoint Measurement, Error and the Rasch Model: A Reply to Michell, and Borsboom and Zand Scholten." *Theory and Psychology* 18 (1): 125–31.

Kyngdon, A. 2008b. "The Rasch Model from the Perspective of the Representational Theory of Measurement." *Theory and Psychology* 18 (1): 89–109.

Latour, B. 1987. *Science in Action: How to Follow Scientists and Engineers through Society.* New York, : Cambridge University Press.

Lattanzio, S., D. S. Burdick, and A. J. Stenner. 2012. *The Ensemble Rasch Model*. Durham, NC: MetaMetrics Paper Series.

Lauer, J. P. 1931. "Étude sur quelques monuments de la IIIe dynastie (pyramide à degrés de Saqqarah)." *Annales du Service des Antiquites de L'Egypte, IFAO* 31 (60): 59.

Lelgemann, D. 2001. *Eratosthenes von Kyrene Und die Messtechnik der Alten Kulturen*. Wiesbaden, Ger.: Chmielorz.

———. 2004. *Recovery of the Ancient System of Length Units*. Berlin: Institute for Geodesy and Geo-Information Technology.

Levy, S. 1994. *Louis Guttman on Theory and Methodology: Selected Writings*. Brookfield, VT: Dartmouth.

Lewin, K. 1951. *Field Theory in Social Science: Selected Theoretical Papers*. New York: Harper & Row.

Lewis, D. 1973. "Causation." *Journal of Philosophy* 70:556–67.

———. 1979. "Counterfactuals Dependence and Time's Arrow." *Nôus* 13:455–76.

Likert, R. 1932. "A Technique for the Measurement of Attitudes." *Archives of Psychology* 140:44–53.

Linacre, J. M. 2002. "Optimizing Rating Scale Category Effectiveness." *Journal of Measurement* 3:85–106.

Linacre, M. 2000. *WINSTEPS*. Chicago: MESA.

Luce, H. D., and L. Narens. 1981. "Axiomatic Measurement Theory." *SIAM-AMS Proceedings* 13:213–35.

Luce, R. D. 1996. "The Ongoing Dialog between Empirical Science and Measurement Theory." *Journal of Mathematical Psychology* 40:78–98.

Luce, R., and J. Tukey. 1964. "Simultaneous Conjoint Measurement: A New Type of Fundamental Measurement." *Journal of Mathematical Psychology* 1:1–27.

Lumsden, J., and J. Ross. 1973. "Validity as Theoretical Equivalence." *Australian Journal of Psychology* 25 (3): 191–97.

Mackowiak, P. A., S. S. Wasserman, and M. M. Levine. 1992. "A Critical Appraisal of 98.6 Degrees F, the Upper Limit of the Normal Body Temperature, and Other Legacies of Carl Reinhold August Wunderlich." *JAMA* 286 (12): 1578–80.

Markus, K. A., and D. Borsboom. 2011. "Reflective Measurement Models, Behavior Domains, and Common Causes." *New Ideas in Psychology.*

Masters, J., and B. Wright. 1984. "The Essential Process in a Family of Measurement Models." *Psychometrika* 49:529–44.

Maudlin, T. 2007. *The Metaphysics within Physics.* Oxford: Oxford University Press.

McArdle, J., K. Grimm, F. Hamagami, R. Bowles, and W. Meredeth. 2009. "Modeling Life-Span Growth Curves of Cognition Using Longitudinal Data with Multiple Samples and Changing Scales of Measurement." *Psychological Methods* 14 (2): 126–49.

McCormick, B. 1964. *Human Engineering.* Warsaw, Pol.: Industrial Design Institute.

Medical Indicators. 2006. [Technical paper]. www.medicalindicators.com.

Meehl, P. E. 1978. "Theoretical Risks and Tabular Asterisks: Sir Karl, Sir Ronald, and the Slow Progress of Soft Psychology." *Journal of Consulting and Clinical Psychology* 46:806–34.

Meek, C., and C. Glymour. 1994. "Conditioning and Intervening." *British Journal for the Philosophy of Science* 45:1001–21.

Menzies, P., and H. Price. 1993. "Causation as a Secondary Quality." *British Journal for the Philosophy of Science* 44:187–203.

MetaMetrics. 1995. *The Lexile Framework*. Research Triangle Park, NC: MetaMetrics.

———. 2000. *The Lexile Framework for Reading*. Durham, NC: MetaMetrics. http://Lexile.com/about/_meta/press/21098b.htm.

Michell, J. 1999. *Measurement in Psychology: A Critical History of a Methodological Concept*. New York: Cambridge University Press.

———. 2003. "Measurement: A Beginner's Guide." *Journal of Applied Measurement* 4 (4): 298–308.

Molenaar, P. C. M. 2004. "A Manifesto on Psychology as Ideographic Science: Bringing the Person Back into Scientific Psychology, This Time Forever." *Measurement: Interdisciplinary Research and Perspectives* 2:201–18.

Monmonier, Mark. 1996. *How to Lie with Maps*. 2nd ed. Chicago: University of Chicago Press.

Nagel, E. 1931. "Measurement." *Erkenntnis* 2:313–33.

Narens, L., and D. R. Luce. 1986. "Measurement: The Theory of Numerical Assignments." *Psychological Bulletin* 99 (2): 166–80.

National Institute of Standards and Technology. "Base Unit Definitions: Second." Accessed September 9, 2016. https://www.physics.nist.gov.

National Research Council of Canada. n.d. "NRC's Cesium Fountain Clock—FCs1." Accessed November 29, 2013.

Newby, V. A., C. Grant, G. Conner, and C. V. Bunderson. 2002. "A Formal Proof That the Rasch Model is a Special Case of Additive Conjoint Measurement." Paper presented at the Eleventh Biennial International Objective Measurement Workshops, New Orleans, LA, April.

Newton, I. 1676. Letter to Robert Hooke, February 15, 1676 [dated as February 5, 1675, using the Julian calendar with March 25 rather than January 1 as New Year's Day, equivalent to February 15, 1676, by Gregorian reckonings]. A facsimile of the original is online at <u>the Digital Library</u>. http://digitallibrary.hsp.org/index.php/Detail/Object/Show/object_id/9285.

————. (1687) 1999. *The Principia: Mathematical Principles of Natural Philosophy.* Translated by I. Bernard Cohen and Anne Whitman. With the introduction *A Guide to Newton's "Principia"* by I. Bernard Cohen. Berkeley: University of California Press.

NexTemp. 2004. Medical Indicators. https://www.medicalindicators.com.

Nichholson, E. 1912. *Men and Measures.* London: Smith, Elder.

Nissen, H. 1988. *Early History of the Ancient Near East.* Chicago: University of Chicago Press.

Norton, J. 2007. "Causation as Folk Science." In *Causation, Physics, and the Constitution of Reality: Russell's Republic Revisited*, 11–44. Oxford: Oxford University Press.

Oates, J. 1986. *Babylon.* London: Thames and Hudson.

Olsen, L. 2002. *Essays on Georg Rasch and His Contributions to Statistics.* Copenhagen, Den.: University of Copenhagen.

Oppenheimer, R. 1956. "Analogy in Science." *American Psychologist* 11:127–35.

Parsons, T. 1937. *The Structure of Social Action.* New York: Free Press.

Payne, S. 1951. *The Art of Asking Questions.* San Francisco: Jossey-Bass.

Pearl, J. (2000) 2009. *Causality: Models, Reasoning, and Inference.* Cambridge: Cambridge University Press. Reprint, New York: Cambridge University Press. Citations refer to first edition.

Pearson, K. 1914. *The Life, Letters and Labours of Francis Galton.* 3 vols. Cambridge: Cambridge University Press.

————. 1920. "Notes on the History of Correlation." *Biometrika* 13:25–45.

Peirce, C. 1940. "Logic as Semiotic: The Theory of Signs." In *The Philosophical Writings of Peirce*, edited by J. Buchler. New York: Dover.

Peirce, C. S. 1998. *The Essential Peirce*. Bloomington: Indiana University Press.

Pellegrino, J. W. 1988. "Mental Models and Mental Tests." In *Test Validity*, edited by H. Wainer and H. Braun. New York: Lawrence Erlbaum.

Perline, R., H. Wainer, and B. D. Wright. 1979. "The Rasch Model as Additive Conjoint Measurement." *Applied Psychological Measurement* 3 (2): 237–55.

Petrie, W. M. F. 1877. *Inductive Metrology*. London: Saunders.

———. 1883. *The Pyramids and Temples of Gizeh*. London: Field and Tuer.

Poincaré, H. 1905. "The Principles of Mathematical Physics." In *Philosophy and Mathematics*, edited by H. Rodgers. Vol. 1 of *Congress of Arts and Science*. Boston: Houghton-Mifflin.

Powell, M. A. 1995. "Metrology and Mathematics in Ancient Mesopotamia." In *Civilizations of the Ancient Near East III*, edited by S. Sasson. New York: Scribner.

Price, H. 1991. "Agency and Probabilistic Causality." *British Journal for the Philosophy of Science* 42:157–76.

Rao, C. S. 1958. "Some Statistical Methods for Comparison of Growth Curves." *Biometrics* 14:1–17.

Rasch, G. (1960) 1980. *Probabilistic Models for Some Intelligence and Attainment Tests*. Reprint, Chicago: University of Chicago Press. Citations refer to the reprinted edition.

———. 1961. "On General Laws and the Meaning of Measurement in Psychology." *Proceedings of the Fourth Berkeley Symposium on Mathematical Statistics and Probability*, 4:321–34. Berkeley: University of California Press.

―――. 1966. "An Individualistic Approach to Item Analysis." In *Readings in Mathematical Social Science*, edited by P. F. Lazarfeld and N. W. Henry. Chicago: Science Research Associates.

―――. 1967. "An Informal Report of Objectivity in Comparisons." In *Psychological Measurement Theory*. Proceedings of the NUFFIC international summer session in science, Het Oude Hof, Den Haag, July 14–28.

―――. 1977. "On Specific Objectivity: An Attempt at Formalizing the Request for Generality and Validity of Scientific Statements." *Danish Yearbook of Philosophy* 14:58–94.

Revised Standard Version of the Bible: RSV. 1952. New York: National Council of the Churches of Christ in the United States of America.

Ricoeur, P. 1977. *The Rule of Metaphor: Multi-disciplinary Studies of the Creation of Meaning in Language*. Translated by R. Czerny. Toronto: University of Toronto Press.

Robbins, W., S. Brody, A. Hogan, C. Jackson, and C. Greene. 1928. *Growth*. New Haven, CT: Yale University Press.

Roche, A. F. 1979. "Secular Trends in Human Growth, Maturation, and Development." *Monograph Social Research in Child Development* 44 (3–4): 1–120.

Rogosa, D. R. 1988. "Myths about Longitudinal Research." In *Methodological Issues in Aging Research*, edited by K. W. Schaie, R. T. Campbell, W. M. Meredith, and S. C. Rawlings, 171–209. New York: Springer.

Rogosa, D. R., D. Brandt, and M. Zimowski. 1982. "A Growth Curve Approach to the Measurement of Change." *Psychological Bulletin* 92 (3): 726–48.

Rogosa, D. R., and J. B. Willett. 1985. "Understanding Correlates of Change by Modeling Individual Differences in Growth." *Psychometrika* 50:203–28.

Rubin, D. 1986. "Comment: Which Ifs Have Causal Answers?" *Journal of the American Statistical Association* 81:961–62.

Salmon, W. 1984. *Scientific Explanation and the Causal Structure of the World.* Princeton, NJ: Princeton University Press.

Schaffer, J. 2000. "Causation by Disconnection." *Philosophy of Science* 67:285–300.

Senecca. 1928. *Quaestiones Naturales.* Bk. 7. Cambridge, MA: Harvard University Press.

Shadish, W., T. Cook, and D. Campbell. 2001. *Experimental and Quasi-Experimental Designs for Generalized Causal Inference.* Chicago: Rand McNally.

Sherry, D. 2011. "Thermoscopes, Thermometers, and the Foundations of Measurement." *Studies in History and Philosophy of Science* 42:509–24.

Shewhart, W. 1931. *Economic Control of Quality of Manufactured Products.* New York: Van Nostrand.

———. 1986. *Statistical Method from the Viewpoint of Quality Control.* New York: Dover.

Shock, N. 1951. "Growth Curves." In *Handbook of Experimental Psychology,* edited by S. Stevens, 330–46. New York: Wiley.

Simon, H. 1969. *The Sciences of the Artificial.* Cambridge, MA: MIT Press.

———. 1989. "Dr. Herbert A. Simon." In *The State of Economic Science,* edited by Werner Sichel, 97–110. Kalamazoo, MI: Upjohn Institute for Employment Research.

Skinner, B. F. 1956. "A Case History in Scientific Method." *American Psychologist* 11:221–33.

Smith, E. V. 2001. "Evidence for the Reliability of Measure and Validity of Measure Interpretation: A Rasch Measurement Perspective." *Journal of Applied Measurement* 2:281–311.

Smith, R. M. 1991a. "The Distributional Properties of Rasch Item Fit Statistics." *Educational and Psychological Measurement* 51:541–65.

———. 1991b. *IPARM: Items and Person Analysis with the Rasch Model.* Chicago: MESA Press.

———. 2000. "Fit Analysis in Latent Trait Measurement Models." *Journal of Applied Measurement* 1 (2): 199–218.

Sosa, E., and M. Tooley. 1993. *Causation.* Oxford: Oxford University Press.

Spirtes, P., C. Glymour, and R. Scheines. 1993. *Causation, Prediction and Search.* New York: Springer-Verlag.

Steckel, R. 2003. "Research Project: A History of Health in Europe from the Late Paleolithic Era to the Present." *Economics & Human Biology* 1:139–42.

Stenner, A. J., and D. S. Burdick. 1997. "The Objective Measurement of Reading Comprehension: In Response to Technical Questions Raised by the California Department of Education Technical Study Group." Unpublished manuscript.

———. 2011. "Can Psychometricians Learn to Think like Physicists?" *Measurement* 9:62–63.

Stenner, A. J., D. S. Burdick, and M. H. Stone. 2008. "Formative and Reflective Models: Can a Rasch Analysis Tell the Difference?" *Rasch Measurement Transactions* 22 (1): 1152–53.

Stenner, A. J., H. Burdick, E. Sanford, and D. S. Burdick. 2006. "How Accurate Are Lexile Text Measures?" *Journal of Applied Measurement* 7 (3): 307–22.

Stenner, A. J., W. Fisher, M. H. Stone, and D. S. Burdick. 2013. "Causal Rasch Models." *Frontiers in Psychology* 13:536. https://doi.org/10.3389/psyg.2013.00536.

Stenner, A. J., and I. Horabin. 1992. "Three Stages of Construct Definition." *Rasch Measurement Transactions* 6 (3): 229.

Stenner, A. J., I. Horabin, and M. Smith. 1989. *The Lexile Scale in Theory and Practice: Final Report*. Washington, DC: MetaMetrics. ERIC Document Reproduction Service (ED 307 577).

Stenner, A. J., and M. Smith. 1982. "Testing Construct Theories." *Perceptual and Motor Skills* 55:415–26.

Stenner, A. J., M. Smith, and D. S. Burdick. 1983. "Toward a Theory of Construct Definition." *Journal of Educational Measurement* 20 (4): 305–16.

Stenner, A. J., and M. H. Stone. 2010. "Generally Objective Measurement of Human Temperature and Reading Ability: Some Corollaries." *Journal of Applied Measurement* 11 (3): 244–52.

Stenner, A. J., M. H. Stone, and D. S. Burdick. 2009. "The Concept of a Measurement Mechanism." *Rasch Measurement Transactions* 23 (2): 1204–6.

———. 2009. "Indexing vs. Measuring." *Rasch Measurement Transactions* 22 (4): 1176–77.

———. 2011. *How to Model and Test for the Mechanisms That Make Measurement Systems Tick*. Paper presented at the Joint International IMEKO Symposium, Jena, Germany, August 31–September 2. urn:nbn:de:gbv:ilm1-2011imeko-027:0.

Stigler, S. 1986. *The History of Statistics*. Cambridge, MA: Harvard University Press.

Stone, M. H. 1995a. "Map of the Variable: WRAT-3." *Rasch Measurement Transactions* 8 (4): 403.

———. 1995b. "Mapping Variables." Paper presented at the Midwest Objective Measurement Society, University of Chicago, Chicago, IL.

———. 1996a. "Data Collecting or Manufacturing?" *Rasch Measurement Transactions* 10:517.

———. 1996b. "An Essay on the Ruler: What Is Measurement?" *Rasch Measurement Transactions* 10:502.

———. 1997. "Steps in Item Construction." *Rasch Measurement Transactions* 11:559.

———. 2000. "Establishing Quality Control in Testing." Paper presented at the Second International Congress on Licensure, Certification and Credentialing of Psychologists, Oslo, Norway, July 20.

———. 2002a. "Data: Collecting or Manufacturing." *Popular Measurement* 4 (1): 15–23.

———. 2002b. *Knox's Cube Test: Revised*. Wood Dale, IL: Stoelting.

———. 2002c. "Quality Control in Testing." *Popular Measurement* 4 (1): 15–23.

Stone, M. H., and A. J. Stenner. 2014. "Comparison Is Key." *Journal of Applied Measurement* 15 (1): 26–39.

———. 2016. "Rasch's Growth Model." Unpublished manuscript (under review for publication).

Stone, M. H., B. Wright, and A. J. Stenner. 1999. "Mapping Variables." *Journal of Outcome Measurement* 3 (4): 306–20.

Stuart, G. E. 1995. "The Timeless Vision of Teotihuacan." *Journal of the National Geographic Society* 88 (6): 2–35.

Swartz, C. W., S. T. Hanlon, A. J. Stenner, and E. L. Childress. 2015. "An Approach to Design-Based Implementation Research to Inform Development of *EdSphere*®: A Brief History about the Evolution of One Personalized Learning Platform." In *Handbook of Research on Computational Tools for Real-World Skill Development*, edited by Y. Rosen, S. Ferrara, and M. Mosharraf. Hershey, PA: IGI Global.

Taagepera, R. 2008. *Making Social Sciences More Scientific: The Need for Predictive Models*. New York: Oxford University Press.

Taylor, B. N., and A. Thompson, eds. 2008. "Appendix 2: Practical Realizations of the Definitions of Some Important Units." In *The International System of Units (SI)*, NIST Special Publication 330, 53ff. Accessed August 25, 2014. http://nvlpubs.nist.gov/nistpubs/Legacy/SP/nistspecialpublication811e2008.pdf.

Thurstone, L. L. 1925. "A Method of Scaling Psychological and Educational Tests." *Journal of Educational Psychology* 16 (7): 433–48.

———. 1926. "The Scoring of Individual Performance." *Journal of Educational Psychology* 17:446–57.

Tufte, E. 1997. *Visual Explanations*. Cheshire, CT: Graphics Press.

Tukey, J. 1969. "Analyzing Data: Sanctification or Detective Work?" *American Psychologist* 24:83–91.

University of Utah. 2010. http://mech.utah.edu/ergo/p.

Vaihinger, H. (1925) 1935. *The Philosophy of "As If": A System of the Theoretical, Practical, and Religious Fictions of Mankind*. New York: Harcourt, Brace. Reprint, London: Kegan Paul.

Waismann, F. 1959. *Mathematical Thinking*. New York: Harper Torchbooks.

———. 1961. *Turning Points in Physics*. New York: Harper Torchbooks.

Waller, J. 2001. "Ideas of Heredity, Reproduction and Eugenics in Britain 1800–1875." *Studies in History and Philosophy of Science Part C* 32 (3): 457–89.

Warwick, D., and D. Lininger. 1975. *The Sample Survey: Theory and Practice*. New York: Mcgraw-Hill.

Encyclopedia Britannica. 1971. Vol. 23. s.v. "Weights and Measures," 371–72. Chicago: Encyclopedia Britannica.

Wilkinson, G. 1993. *Administration Manual: WRAT-R*. Wilmington, DE: Wide Range.

———. 1993. *Manual for Wide Range Achievement Test-3*. Wilmington, DE: Wide Range.

Wilson, M. 2004. *Constructing Measures: An Item Response Modeling Approach*. Mahwah, NJ: Lawrence Erlbaum.

Wilson, M., X. Zheng, and L. McGuire. 2012. "Formulating Latent Growth Using an Explanatory Item Response Model Approach." *Journal of Applied Measurement* 13 (1): 1–22.

Wittgenstein, L. 1922. *Tractatus Logico-Philosophicus*. London: Routledge & Kegan Paul.

———. 1958. *Philosophical Investigations*. New York: Macmillan.

Woodward, J. 1984. *A Theory of Singular Causal Explanation*. Ithaca, NY: Cornell University Press.

———. 1997. "Explanation, Invariance, and Intervention." *PSA* 1996 (2): 26–41.

———. 2000. "Explanation and Invariance in the Special Sciences." *British Journal for the Philosophy of Science* 51:197–254.

———. 2003. *Making Things Happen*. New York: Oxford University Press.

———. 2007. "Causation with a Human Face." In *Causation, Physics, and the Constitution of Reality: Russell's Republic Revisited*, 66–105. Oxford: Oxford University Press.

———. 2011. "Data and Phenomena: A Restatement and Defense." *Synthese* 182 (1): 165–79.

Woodward, J., and C. Hitchcock. 2003. "Explanatory Generalizations." Pt 1, "A Counterfactual Account." *Nôus* 37:1–24.

Wright, B. D. 1977. "Solving Measurement Problems with the Rasch Model." *Journal of Educational Measurement* 14:97–116.

———. 1999. "Fundamental Measurement for Psychology." In *The New Rules of Measurement: What Every Educator and Psychologist*

Should Know, edited by S. Embretson and S. Hershberger, 65–104. Hillsdale, NJ: Lawrence Erlbaum.

Wright, B. D., and M. Linacre. 1987. "Rasch Model Derived from Objectivity." *Rasch Measurement Transactions* 1 (1): 5–7.

Wright, B. D., and G. N. Masters. 1982. *Rating Scale Analysis.* Chicago: MESA.

Wright, B. D., and M. H. Stone. 1979. *Best Test Design.* Chicago: MESA.

———. 1999. *Measurement Essentials.* Wilmington, DE: Wide Range.

———. 2003. *Managing Observations, Inventing Constructs, Crafting Yardsticks, Examining Fit.* Chicago: Phaneron Press.

von Wright, G. 1971. *Explanation and Understanding.* Ithaca, NY: Cornell University Press.

Zubin, J. 1955. "Clinical vs. Actuarial Prediction: A Pseudo-problem." In *Proceedings of the 1955 Invitational Conference on Testing Problems,* 107–28. Princeton, NJ: Educational Testing Service.

———. 1955. *Experimental Abnormal Psychology.* New York: Columbia University Store. Mimeographed.

Zupko, E. 1990. *Revolution in Measurement: Western European Weights and Measures Since the Age of Science.* Philadelphia: American Philosophical Society.

The authors recommend special attention to the issues of growth in papers by Gary Williamson, Ph.D.

Williamson, G. L. (2015). Measuring academic growth contextualizes text complexity. Pensamiento Educativo: Revista de Investigación Educacional Latinoamericana, 52(2), 98-118. http://pensamientoeducativo.uc.cl/index.php/pel/article/view/731/1464

Williamson, G. L. (2016). Novel interpretations of academic growth. Journal of Applied Educational and Policy Research, 2(2), 15-35. https://journals.uncc.edu/jaepr/article/view/509

Williamson, G. L. (2018). Exploring reading and mathematics growth through psychometric innovations applied to longitudinal data. Cogent Education, 5(1): 1464424. https://www.tandfonline.com/doi/full/10.1080/2331186X.2018.1464424

INDEX

f denotes figure; *t* denotes table

A

absolute measurement, 61
absolute scale, 61
*An Absolute Scale of Binet Test
    Questions* (Thurstone), 56–
    57, 57*f*
abstraction
    pathway to, 39
    role of in evolution of human
        science instrumentation,
        33–34
accuracy, 14, 25, 48, 52, 53, 56, 177,
    212, 217, 234, 295
achievement
    abstract variables for
        determining growth
        of, 261
    achievement tests, 189
    measurement of, 196–198
    of objectivity, 41–42
    reading achievement, 26
    Wide Range Achievement Test
        (WRAT), 58, 58*f*, 196–198
Adams, John Crouch, 301
additive conjoint measurement,
    103, 162–165, 221
The Adoration of the Magi (painting
    by Dondone), 300
Aesop, 32

age parameter, 263, 270, 271
agency theory, 112–118
agent probabilities, 114–115
Alexander, H., 178
analogy
    according to Oppenheimer, 54
    map as, 53–55
    use of, 14–15, 16, 17, 18, 36, 47,
        91, 100, 139, 142, 255, 299
Analogy in Science
    (Oppenheimer), 14
Andrich, D., 4, 63, 88, 147, 262,
    264–265, 286
anomalies, 6, 63, 66, 91
ANSUR, 241
arithmetic
    manipulation of things by, 293
    "measure" as abstract number
        with which arithmetic
        can be done, 206
    measurement of arithmetic
        computation, 58, 189
    as pursuing uniform
        quanta, 14
    in substantive scale
        construction, 195–196
arrow, as symbol, 100
artifacts, 33, 73, 77, 80, 94, 97–98,
    187, 248, 251, 307, 309

CRM (causal Rasch model). *See*
  causal Rasch model (CRM)
cross-level inferences, 9
CTT (classical test theory), 217
cubit
  causal Rasch model versus
    descriptive Rasch model,
    256–258
  constructing measures,
    250–256
  cubit dimensions from
    Berriman, 233*t*
  cubit frequency by inches for
    348 subjects, 237*f*
  derivation of word, 223
  descriptive statistics for A. E.
    Berriman's table, 234*t*
  discussion, 242–243
  Egypt, 229
  elbow-fingertip length
    percentile distribution in
    millimeters, 241*t*
  forearm percentiles for
    unidentified British
    population, 240*t*
  frequency of left cubit measure
    by inches, 237*t*
  Greek and Roman
    comparisons, 230
  Greek/Roman periods, 230–231
  human cubit, 236–241
  human dimensions relative to
    six-foot male, 235*t*
  introduction, 223–228
  KCT and empirical
    demonstration, 259–260
  measuring mechanism,
    246–250
  Middle East names and
    dimensions for cubit and
    related measures, 233*t*

millimeters and inches of left
  cubit, 238*t*
other Near East dimensions,
  231–235
plot of left cubit to stature, 239*f*
sensation to gradation, 250
Thales and Rasch, 243–244
three-dimensional view of
  Galton's data, 239*f*
time, 244–246

D

Da Vinci, Leonardo, 234–235, 239
Dantzig, T., 38, 53
data
  and models, 203–204
  theory and, 299–306
Data: Collecting or Manufacturing
  (Stone), 255
data manufacturing RMT, 184–185
David ben Zimra, 227
Dawes, R. M., 162
Dedekind, Richard, 291–292, 293
Deming, W. Edwards, 176, 178
dependent variable, 7, 67, 78, 91,
  100, 101, 259, 263, 281
depression inventory, 200
descriptive Rasch model (DRM)
  versus causal Rasch model, 13,
    84, 88–92, 256–258
  versus causation, 63–71
deterministic, 123, 129, 161, 264
developmental science, 75–77
difficulty variance, 186
dimensions, relative lengths of
  four common dimensions,
  224*t*
directed graphs, and
  manipulationist theories of
  causation, 120–124
discontinuous (discrete), 291

indices, 18, 62, 150, 175, 180, 181, 218, 219, 220, 221

individual measurement, 68, 82, 92, 288

individual-centered models/ measures, 9, 72, 80–84, 85, 96, 286–291

information, 22, 42, 52, 55, 60, 61, 62, 68, 93, 120, 123, 124, 159, 165, 180, 198, 199, 200, 202, 260, 266, 267, 274, 299

instructional theory, connecting psychometrics to, for language development, 71–75, 96–99

instruments
    engineering of, 5, 65, 84, 89, 90, 290, 291
    as islands unto themselves, 2–3
    parallel instruments, 87–88
    validity of, 311

intellectual development, according to Binet, 108

intelligence, according to Binet, 107–108

International Practical Temperature Scale, 18

International Temperature Scale of 1990 (ITS-90), 18–19

The Interpreter's Bible, 226

interventionist accounts, 125, 131–138

interventionist formulation/ reformulation, 111, 112

interventionist framework/ approach, 109, 121, 122, 127–128, 129, 130

interventions
    and counterfactuals, 129–130
    described, 123
    Pearl's characterization of, 122

possible and impossible ones, 130–131

that do not involved human action, 128–129

use of term, 119

invariance, 4, 5, 10, 17, 35–36, 64, 66, 68, 84, 88, 91, 92, 102, 103, 221, 250, 289, 311. *See also* dual invariance requirements; multiple invariance; scale invariance

invariant comparisons, 5

Inventing Temperature (Chang), 75

inventions
    temperature, 248, 250
    time, 248

IRT models, 3, 13

item analysis, xii, 32, 261

item banking, 185–187

item construction, 32, 189, 191

item difficulty, 27, 56, 58, 168, 170, 180, 185, 186, 193, 198, 210, 211, 259, 260

item parameter, 3, 10, 12, 41, 44, 87, 101, 106, 163, 166, 210, 259

Item Response Theory, 64, 88

Item Separation Reliability (ISR), 209

item specification, versus item banking, 186–187

item stimuli, dissection and manipulation of, 3

item try-outs, 194–195

ITS-90 (International Temperature Scale of 1990), 18–19

J

Jackson, R., 178, 261

Joad, C., 305

John of Salisbury, 75

Johnson, R. C., 240

manipulability approach, 111, 112, 135

manipulability theory, 109, 110–111, 112, 117, 119, 124–125, 128, 129, 130, 133, 134, 135, 136

manufacturing, 77, 78, 83, 84, 86, 87, 174, 175, 176, 203–204, 247, 251, 254, 255, 289, 290, 291. *See also* data manufacturing RMT; Lean Manufacturing Strategy

mapping
    flowchart of, 63
    origins of, 45–52
    in substantive scale construction, 201–203
    of variables, 45–84

maps
    as analogies, 53–55
    examples of early maps, 45–52, 46*f*, 47*f*, 48*f*, 49*f*, 50*f*, 51*f*
    examples of humorous maps, 54–55
    graphs as, 53
    in psychometrics, 55–61

Markus, K. A., 13

McCormick, B., 241

measure, defined, 206

measurement
    absolute measurement, 61
    of achievement, 196–198
    additive conjoint measurement, 103, 162–165, 221
    Binet and constructive measurement, 107–108
    concatenating sticks and, 139–142
    dialogue of with empirical science, 37

educational measurement, 86–87

efficiency of linear measurement, 21

as embodying construct theory, 34, 185, 260

goal of, xi

of height, 82–83, 211–212, 288–289

individual measurement, 68, 82, 92, 288

as made by analogy, 14, 53. *See also* analogy

new paradigm for in social sciences, 1–2

of quality control, 176–181

of reading, 289. *See also* reading ability, measurement of

of reading ability, 8, 22, 26, 30, 74, 85, 99, 106, 281, 282, 286

in social science as compared to physical sciences, 2, 9

of spelling ability, 191–194, 294–295

of temperature, 6, 8, 81, 90, 99, 142–143

of text complexity, 12, 68, 92

as theoretical claim, 104

thermometer as good paradigm for, 18–19

of time, 14

uncertainty of, 5, 84, 90

of weight, 2, 5, 6, 33, 54, 65, 66, 84, 89–90, 264, 290

measurement error, 73, 84, 97, 179, 208, 274, 309

Measurement Essentials (Wright and Stone), 139

measurement instrument, explanation of working of, 7

precision, xi, 33, 56, 174, 177–181, 184, 187, 203, 208, 219, 245, 246, 250

predictions, 102, 104, 178

predictive control, as hallmark of success, 34

prescriptive, 4, 63, 88, 254, 257–258. *See also* doubly prescriptive Rasch model

Price, Huw, 34, 109, 111, 114–118, 121, 123, 125, 128

Principia (Newton), 245, 300, 304

Probabilistic Models for Some Intelligence and Attainment Tests, 80, 286

probability, 79, 114, 123, 135, 136, 165, 168, 217, 267, 274, 275. *See also* agent probabilities; conditional probability; unconditional probability

probability models, 7, 67, 92, 104–105, 247

PSR (Person Separation Reliability), 208

psychogenetic method, 108

psychometrics
atheoretical psychometrics, 104
connecting of to instructional theory for language development, 71–75, 96–98, 97f, 98f
focus of, 1
map topography as applicable to, 52
maps in, 55–61

Q

qualitative, versus quantitative, 291

qualitative observations, 15, 147, 156, 255

qualitative validity, 209

quality, steps from quality to quantity, 165–172

quality control, 34, 63, 77, 78, 83, 84, 86, 142, 158, 161, 175, 176–181, 184, 185, 186, 203–204, 253, 254, 255, 258, 290, 297, 307

quantitative attribute, 11, 12, 13

quantitative hypothesis, 72, 97

quantitivity hypothesis, 10, 11, 12

quantity
axioms of, 140
from ordinality to quantity, 299
steps from quality to quantity, 165–172

R

Rao, C. R., 264, 280, 286

Rasch, Georg. *See also* causal Rasch model (CRM)
on achievement of objectivity, 41–42
application of models of, 83
association of with item analysis, xii
on constructing experimental comparisons, 156–162
dichotomous model of, 7, 27, 67, 91, 101, 165, 204–205
on discussing models in classical physics, 249
estimation programs of, 44
goal of, 15, 255
growth model of, 261–311
influence on building of maps for major dimension of human behavior, 61
influence on thinking of by Zubin, 81
mapping technology of, 60, 60f

as not enjoying commonly
adhered to construction
definition, 6
Test of English Reading Ability
(TERA), 311
reading construct, evidence for
reality of, 6
reading corpus, 70, 94
Reading Proficiency Scale (RPS), 61
reading test, described, 26
reading theory, 8, 75, 99, 106, 213.
See also Lexile theory
reference, frame of. *See* frame of
reference
referential, 273
reflective model, 215, 216, 217, 218
relationship
between additive conjoint
measurement and Rasch
model, 163
causal relationships. *See* causal
relationships, 34
in combined gas law, 31, 212
of cubit to stature, 239
data-model relationship, 4,
63, 88
establishment of by employing
prediction as critical
determinant, 308
functional relationship, 35–
36, 261
graphs of functions as maps
showing, 53
between KCTR substantive
theory and specification
equation to observed
data, 170–171
between latent variable
and measurement
outcome, 22

between measurement
outcome and its
measure, 23
of Mohs scale to Vickers, 298
physical relationship, 296
between raw counts, 206
in reading law, 31, 213
significance test as evaluating
relationship between
dependent variable
and independent
variable(s), 78
of theory and data, 304, 305
theory as based upon, 33
theory/data relationship
utilizing
manipulation, 257
between things and
numbers, 292
between two or more
statistical variables, 258
reliability
perspectives on, 78
in substantive scale
construction, 208–209
as whole instrument, 1
reliable variance, 187
representational measurement
theory (RPM), 38
response alternative, 189, 190, 205
response format, 191, 193, 199, 200,
201, 204
Ricoeur, P., 161
RMT 22:1, 218
Robbins, W., 261
Roche, A. F., 242
Rogosa, D. R., 285
Rose, Steven, 261
Ross, J., xii, 307, 308, 310, 311
RPM (representational
measurement theory), 38

Stenner, A. J., 71, 96, 101, 164, 171, 218, 246, 248, 251, 257
Stevens, S., 35–36, 293
stick ruler, 139
Stigler, S., 236, 239, 240
stochastic framework/model, 155, 161, 204, 213, 286
stochastic subject interpretation, 22
Stone, M. H., 101, 164, 178, 187, 200, 201, 218, 248, 251, 255, 259
Strabo, 48, 230
structural equation modeling (SEM), 215, 217, 266, 274, 275
structural equations, 34, 110, 120–124, 263, 265, 266
structural theory, use of term, 308
The Structure of Scientific Revolutions (Kuhn), 75–76
Stuart, G. E., 46–47
substantive scale construction
    achievement, 196–198
    arithmetic, 195–196
    attitude scales, 189
    bipolar response alternatives, 189–190
    building yardsticks, 101
    combined gas law and Rasch Reading Law, 212–215
    counts and measures, 207–208, 207f
    data and models, 203–204
    formative and reflective models, 215–218
    indexing versus measuring, 218–221
    item try-outs, 194–195
    linking, 190–191
    mapping, 201–203
    Rasch models, 204–206
    rating scale formats, 200–201
    rating scales, 198–199

    reliability, 208–209
    response format, 199
    sample issues, 210–211
    from scores to measures, 206–207
    spelling, 191–194
    validity, 209–210
substantive theory, xi, 1, 2, 3, 5, 6, 7, 10, 22, 30, 64, 66, 67, 68, 72, 77, 81, 83–84, 85, 86, 87, 89, 90, 91, 92, 97, 100, 101, 103, 104, 107, 108, 150, 168, 170, 172, 214, 220, 221, 247, 248, 250, 251, 253, 254, 256, 287, 289, 290, 299, 307, 308
Swartz, C. W., 71, 96
Synopsis of Cometary Astronomy, 300
systematic variance, 307, 308, 310, 311
Systeme International des Unites (SI), 296

T

Taagepera, Rein, 147, 309, 310
targeting, 175, 179, 180, 183
Temperance (Pieter Bruegel the Elder), 13, 15f
temperature
    Celsius temperature, 15–17
    as compared and contrasted to reader ability, 21–44
    defined, 18
    Fahrenheit temperature, 15–17
    human body temperature, 23–26
    invention of, 248, 250
    meaning of, 15
    measurement of, 6, 8, 81, 90, 99, 142–143
    "normal" temperature, 24
Teotihuacan, map of, 46–47

Wolpe, J., 200
Woodward, J., 109, 111, 124
WRAT-3, 58, 58*f*, 198
Wright, Ben, 61, 139, 162, 163, 178,
    200, 217
Wright maps, 3
writing, why we write the way we
    do, 306–307
Wunderlich, Carl, 24

Y

yardstick, building of, 191

Z

Zimowski, M., 285
Zubin, J., 80–81, 286, 287